Owning the Technical Baseline for Acquisition Programs in the U.S. Air Force

A WORKSHOP REPORT

Committee on Owning the Technical Baseline in the U.S. Air Force: A Workshop

Air Force Studies Board

Division on Engineering and Physical Sciences

NATIONAL RESEARCH COUNCIL
OF THE NATIONAL ACADEMIES

THE NATIONAL ACADEMIES PRESS
Washington, D.C.
www.nap.edu

THE NATIONAL ACADEMIES PRESS **500 Fifth Street, NW** **Washington, DC 20001**

NOTICE: The project that is the subject of this report was approved by the Governing Board of the National Research Council, whose members are drawn from the councils of the National Academy of Sciences, the National Academy of Engineering, and the Institute of Medicine. The members of the committee responsible for the report were chosen for their special competences and with regard for appropriate balance.

This is a report of work supported by FA9550-14-1-0127 between the U.S. Air Force and the National Academy of Sciences. Any opinions, findings, or conclusions expressed in this publication are those of the author(s) and do not necessarily reflect the view of the organizations or agencies that provided support for the project.

International Standard Book Number-13: 978-0-309-37431-6
International Standard Book Number-10: 0-309-37431-6

Limited copies of this report are available from:

Air Force Studies Board
National Research Council
500 Fifth Street, NW
Washington, DC 20001
(202) 334-3111

Additional copies are available from:

The National Academies Press
500 Fifth Street, NW
Keck 360
Washington, DC 20001
(800) 624-6242 or (202) 334-3313
http://www.nap.edu

Printed in the United States of America

THE NATIONAL ACADEMIES

Advisers to the Nation on Science, Engineering, and Medicine

The **National Academy of Sciences** is a private, nonprofit, self-perpetuating society of distinguished scholars engaged in scientific and engineering research, dedicated to the furtherance of science and technology and to their use for the general welfare. Upon the authority of the charter granted to it by the Congress in 1863, the Academy has a mandate that requires it to advise the federal government on scientific and technical matters. Dr. Ralph J. Cicerone is president of the National Academy of Sciences.

The **National Academy of Engineering** was established in 1964, under the charter of the National Academy of Sciences, as a parallel organization of outstanding engineers. It is autonomous in its administration and in the selection of its members, sharing with the National Academy of Sciences the responsibility for advising the federal government. The National Academy of Engineering also sponsors engineering programs aimed at meeting national needs, encourages education and research, and recognizes the superior achievements of engineers. Dr. C. D. Mote, Jr., is president of the National Academy of Engineering.

The **Institute of Medicine** was established in 1970 by the National Academy of Sciences to secure the services of eminent members of appropriate professions in the examination of policy matters pertaining to the health of the public. The Institute acts under the responsibility given to the National Academy of Sciences by its congressional charter to be an adviser to the federal government and, upon its own initiative, to identify issues of medical care, research, and education. Dr. Victor J. Dzau is president of the Institute of Medicine.

The **National Research Council** was organized by the National Academy of Sciences in 1916 to associate the broad community of science and technology with the Academy's purposes of furthering knowledge and advising the federal government. Functioning in accordance with general policies determined by the Academy, the Council has become the principal operating agency of both the National Academy of Sciences and the National Academy of Engineering in providing services to the government, the public, and the scientific and engineering communities. The Council is administered jointly by both Academies and the Institute of Medicine. Dr. Ralph J. Cicerone and Dr. C. D. Mote, Jr., are chair and vice chair, respectively, of the National Research Council.

www.nationalacademies.org

COMMITTEE ON OWNING THE TECHNICAL BASELINE IN THE U.S. AIR FORCE: A WORKSHOP

HENRY A. "TREY" OBERING III, Booz Allen Hamilton, *Chair*
CLAUDE M. BOLTON, JR., Defense Acquisition University
MICHAEL D. GRIFFIN, Schafer Corporation
TERRY J. JAGGERS, Decisive Analytics Corporation
GARY A. KYLE, Persistent Agility, Inc.
JON S. OGG, AeroVise, LLC
SUE C. PAYTON, SCI Aerospace, Inc.
RICHARD T. ROCA, Johns Hopkins University Applied Physics Laboratory

Staff

JOAN FULLER, Director, Air Force Studies Board
CARTER W. FORD, Program Officer
DIONNA C. ALI, Research Assistant
GREGORY EYRING, Senior Program Officer
MARGUERITE E. SCHNEIDER, Administrative Coordinator

Acknowledgment of Reviewers

This report has been reviewed in draft form by individuals chosen for diverse perspectives and technical expertise, in accordance with procedures approved by the National Research Council's (NRC's) Report Review Committee. The purpose of this independent review is to provide candid and critical comments that will assist the institution in making its published report as sound as possible and to ensure that the report meets institutional standards for objectivity, evidence, and responsiveness to the study charge. The review comments and draft manuscript remain confidential to protect the integrity of the deliberative process. We wish to thank the following individuals for their review of this report:

Naim Awaad, Space and Missile Systems Center, Air Force Space Command,
R. Stephen Berry, University of Chicago,
Gordon R. England, E6 Partners, LLC,
Millard S. Firebaugh, University of Maryland, College Park,
Paul G. Gaffney II, Monmouth University, and
Kathleen K. Paige, Mark India LLC.

Although the reviewers listed above have provided many constructive comments and suggestions, they were not asked to endorse the views presented at the workshop, nor did they see the final draft of the workshop report before its release. The review of this workshop report was overseen by Chris G. Whipple, ENVIRON (retired). Appointed by the NRC, he was responsible for making certain that an independent examination of this workshop report was carried out in accordance with institutional procedures and that all review comments were carefully considered. Responsibility for the final content of this report rests entirely with the committee and the institution.

Contents

Acronyms

A&AS	advisory and assistance services
AFLCMC	Air Force Life Cycle Management Center
AFMC	Air Force Materiel Command
AFRL	Air Force Research Laboratory
APT	Advanced Pilot Training
DCGS	Distributed Common Ground Systems
DoD	Department of Defense
ECSS	Expeditionary Combat Support System
EPASS	Engineering, Professional, and Administrative Support Services
FAB-T	Family of Advanced Beyond-Line-of-Sight Terminals
FFRDC	federally funded research and development center
FoS	Family of Systems
ISEC	ISR Systems Engineering Center
ISR	intelligence, surveillance, and reconnaissance
Lab Demo	laboratory demonstration
LPTA	lowest price technically acceptable
LRSO	Long-Range Stand Off
NRC	National Research Council
O&M	operations and maintenance
OTB	owning the technical baseline
PEO	program executive officer

PM	program manager
PO	program office
RDT&E	research, development, test, and evaluation
S&E	science and engineering
SAE	service acquisition executive
SAF/AA	Secretary of the Air Force, Administrative Assistant
SAF/AQ	Assistant Secretary of the Air Force Acquisition
SAF/AQR	Deputy Assistant Secretary of the Air Force for Science, Technology, and Engineering
SAF/FM	Assistant Secretary of the Air Force for Financial Management
SAF/PAO	Secretary of the Air Force, Public Affairs Office
SMC	Space Missile Command
SPO	system program office
TSPR	total system performance responsibility
UARC	university affiliated research center
UMD	Unit Manpower Document
W/G/S/F	wings, groups, squadrons, and flights

Overview

CONTEXT FOR THE WORKSHOP

The ability to be a "smart buyer" is dependent on unrestricted access to and understanding of a program's technical baseline combined with a science and engineering workforce that has the "know-how"—that is, knowledge, skills, and abilities—to ensure program success.[1,2] A number of observers have argued that for years, beginning in the mid-1990s, the U.S. Air Force ceded control, active oversight, and in-depth understanding of the technical baselines for weapon systems to defense prime contractors with negative consequences that included (1) loss of ability to perform independent technical analysis, (2) loss of ability to validate defense contractor technical decisions and conclusions, (3) atrophy of the engineering workforce competency, (4) decrease in the ability to attract and retain top engineering talent due to hands-off engineering, (5) decreased ability to control costs, and (6) a reluctance by industry to share detailed, proprietary technical data for fear of transfer to competitors. In addition, others have asserted that pressures imposed by the Budget Control Act of 2011 (P.L. 112-25, also referred to as sequestration) will exacerbate these challenges in the coming years, which will necessitate multifaceted and concerted efforts by the Air Force and Department of Defense (DoD) to look for efficiencies while simultaneously increasing effectiveness. The Better Buying Power initiatives—launched by the Under Secretary of Defense for Acquisition, Technology, and Logistics in 2010, along with earlier provisions under the Weapon System Acquisition Reform Act and the Defense Acquisition Workforce Improvement Act, to improve DoD's ability to technically manage complex weapon systems throughout their life cycle in collaboration with industry contractors—are intended to address these goals.[3,4] The question is, what is the appropriate balance between the notion of government as system integrator on one extreme and the total system performance responsibility approach—in which government's role in the management of the system is minimized—on the other?

[1] The Air Force defines *technical baseline* as "Data and information that provides the program office knowledge to establish, trade-off, verify, change, accept, and sustain functional capabilities, design characteristics, affordability, schedule, and quantified performance parameters at the chosen level of the system hierarchy." *Ownership of the technical baseline* is defined as "The government applies technical baseline knowledge to be an informed decision maker." SOURCE: Bud Boulter, SAF/AQRE, and David Crawford, MITRE, "OTB Implementation," presentation to the committee on November 4, 2014.

[2] *Technical baseline* is distinct from *tech base,* the latter term more often associated with Department of Defense basic research (6.1 and 6.2).

[3] See Defense Acquisition University, https://acc.dau.mil/wsara.

[4] See U.S. Code, Title 10, Chapter 87, Defense Acquisition Workforce, text contains those laws in effect on September 18, 2013, available at http://www.acq.osd.mil/se/docs/dawia.pdf.

In this context, the Assistant Secretary of the Air Force for Acquisition requested that the Air Force Studies Board of the National Academies' National Research Council (NRC) undertake a workshop with the following terms of reference: (1) identify the essential elements of the technical baseline that would benefit from realignment under Air Force or government ownership, and the value to the Air Force of regaining ownership under its design capture process of the future; (2) identify the barriers that must be addressed for the Air Force to regain technical baseline control to include workforce, policy and process, funding, culture, contracts, and other factors; and (3) provide terms of reference for a possible follow-on study to explore the issues and make recommendations required to implement and institutionalize the technical baseline concept, and possibly prototype the concept on a demonstration program for lessons learned. Figure O-1 provides a graphical depiction of how the workshop relates to current Air Force and DoD initiatives.

A committee, the Committee on Owning the Technical Baseline in the U.S. Air Force: A Workshop, appointed by the NRC in October 2014 under the auspices of the Air Force Studies Board (see Appendix A), planned and participated in the workshop and prepared the report. *This report summarizes individual committee members' observations, based on presentations that they received from 35 DoD program managers, current and senior DoD leaders, and program contract managers and also highlights a select set of these presentations. While the committee is responsible for the overall quality and accuracy of the report as a record of what transpired at the workshop, the views contained in this section and in the rest of the report are not necessarily those of all workshop participants, the committee, or the NRC. The committee was not asked to reach consensus on conclusions, findings, or recommendations.*

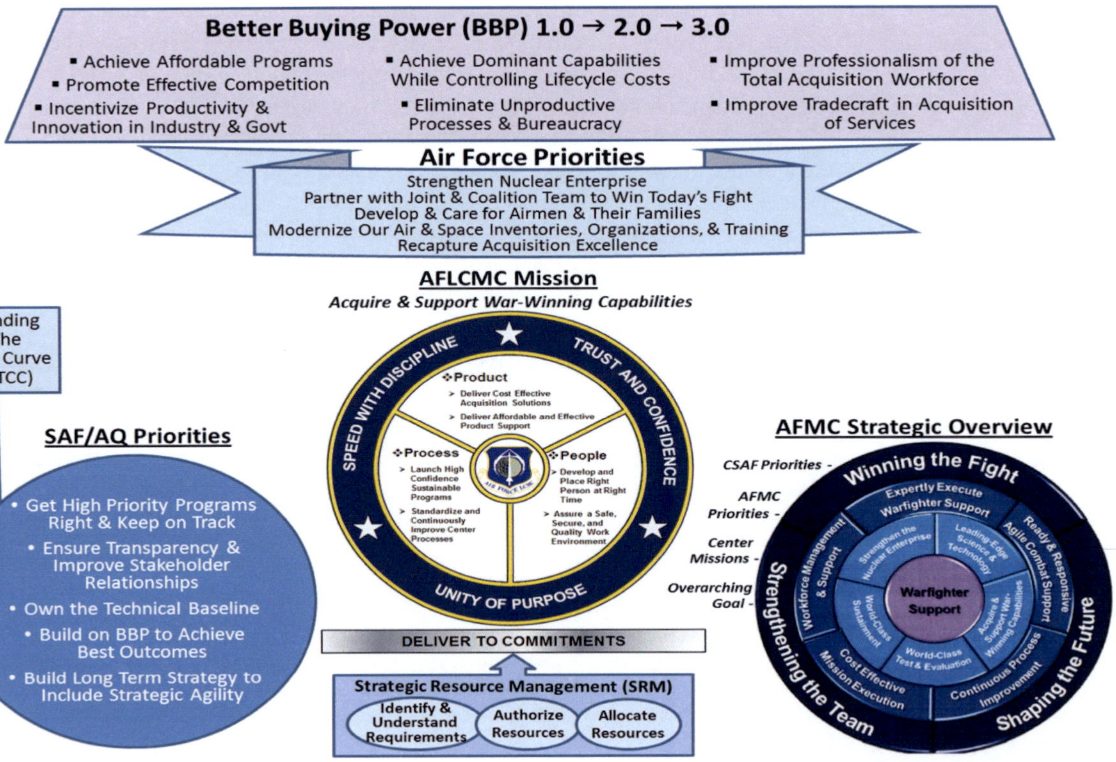

FIGURE O-1 "Owning the Technical Baseline" roles and responsibilities as part of an overall Department of Defense acquisition strategy. SOURCE: Lt Gen John Thompson, Commander, Air Force Life Cycle Management Center. Approved for release by SAF/PAO on December 18, 2014.

Three workshop sessions were held on November 4-5, 2014 (Washington, D.C.); December 17-18, 2014 (Dayton, Ohio); and January 20-22, 2015 (Washington, D.C.) (see Appendix B). DoD program managers, current and senior DoD leaders, and program contract managers made presentations at the workshop. DoD program managers were encouraged to respond to the following questions, but no attempt was made by the organizing committee, workshop participants, or staff to evaluate or assess the health or likelihood of success of any of the programs discussed:

1. How would you describe the health of your program?
2. What constitutes the technical baseline for your program and what is it based on?
3. To what level does the government manage the technical baseline, and how does your contract enable you to do so?
4. What mechanisms/processes are in place to manage/control the technical baseline?
5. Do you feel properly resourced to manage your technical baseline?

Senior DoD leaders were asked to respond to the following questions:

1. How do you define a program's technical baseline?
2. What policy guidance or instruction has been provided to the program executive officers (PEOs) and program managers (PMs) regarding ownership of the technical baseline?
3. To what extent are you willing to fund the programs to be able to take control and manage their technical baselines?
4. Do you believe they are already adequately resourced to manage the technical baseline?

The scope of the workshop included not only current and future Air Force and DoD programs, but also past programs, such as the Expeditionary Combat Support System, where the workshop participants could potentially glean lessons learned and also practices that could be applied to current and future programs. Presentations at the workshop served as useful points of discussion throughout the workshop and the basis for the discussions summarized below.

SUMMARY OF RECURRING TOPICS ARISING DURING THE WORKSHOP

During the course of the three 2-day workshop sessions, several recurring topics emerged as a result of various presentations and resulting dialogue among the participants. These issues are summarized below along with the names of the participants who identified the common topic. Details underlying each topic are found in Chapter 2.

Programs: According to several workshop participants (Terry Jaggers, Trey Obering, Richard Roca), the most important issue with respect to managing the technical baseline of a weapon system is for the PEO and PM to be able to oversee and manage the baseline with accountability, authority, and responsibility.

Leadership and Culture: In the opinion of at least two participants (Michael Griffin, Trey Obering), it is essential that senior Air Force leadership make it clear to all functional leaders supporting acquisition that the Air Force highly values technically trained and competent acquisition and engineering personnel.

Workforce: In the view of three participants (Terry Jaggers, Jon Ogg, Sue Payton), continuity, longevity, and mentoring in the engineering and technical fields, including a succession pipeline, are crucial for success of a program. They argued that the Air Force needs to implement a formal, robust, and credible training and mentoring program to (1) transfer knowledge to upcoming acquisition professionals and (2) develop demonstrable business acumen in the acquisition workforce.

Funding: According to several participants (Claude Bolton, Jr., Jon Ogg, Sue Payton), the limitations of operations and maintenance (O&M) funding and the inability to use research, development, test, and evaluation (RDT&E) funding for hiring, retaining, and training the technical acquisition workforce create barriers to success. In their view, this lack of adequate and timely funding limits the ability of acquisition-center functional leads from shaping the workforce to meet the demands for knowledgeable and experienced technical talent.

Contracting: In the opinion of two participants (Terry Jaggers, Gary Kyle), it is essential that contracts reflect the proper level of government technical and business engagement to include oversight, insight, data rights, and intellectual property consistent with the program's life cycle acquisition strategy. This should start at the earliest phases of the program and include a maintenance or sustainment approach for owning the technical baseline over the life cycle of the program.

1

Why Owning the Technical Baseline for Acquisition Programs Is Important to the U.S. Air Force

PROBLEM STATEMENT

The U.S. Air Force has experienced many acquisition program failures—cost overruns, schedule delays, system performance problems, and sustainability concerns—over program lifetimes. A key contributing factor is the lack of sufficient technical knowledge within the Air Force concerning the systems being acquired to ensure success. This workshop was charged with examining this issue.

DEFINITIONS

In addressing Task 1 of the terms of reference (see the Overview, p. 2), the committee recognized early on that it would be important to clearly define the term "technical baseline," because it appears to mean different things to different people and programs. According to Richard Roca, director emeritus of the Johns Hopkins University Applied Physics Laboratory, the technical baseline has many individual elements, including system designs, interface controls, system models, development and operational data, architecture, cost data, performance data, risk analyses, operational processes, and training requirements. However, rather than focus on the individual elements, for purposes of this workshop the committee chose to focus on the uses to which the information could be put, defining the technical baseline and, similarly, defining "owning the technical baseline."

> *Technical baseline:* Data and information that provides the program office knowledge to establish, trade-off, verify, change, accept, and sustain functional capabilities, design characteristics, affordability, schedule, and quantified performance parameters at the chosen level of the system hierarchy.

> *Owning the technical baseline:* Air Force program managers and personnel have sufficient technical knowledge of their engineering development programs to ensure program success by making informed, timely, and independent decisions to manage cost, schedule, and performance risk while ensuring disciplined program execution. Owning the technical baseline allows the Air Force to respond knowledgeably and have minimal disruption to mission success.

In this sense, "owning" does not necessarily mean buying the technical data or other necessary information. According to Claude Bolton, Jr., executive-in-residence at the Defense Acquisition University, the term does mean, in part, having ready access to the data and information necessary for engaging in critical assessments and analyses, and the government being smart enough to understand the baseline completely and know what parts of the baseline it needs to validate the contractors' product. In the opinion of Michael Griffin, former administrator of the National Aeronautics and Space Administration, it is critical that the government have sufficient technical information about—and the competence to understand—the critical cost, schedule, and performance trade-offs that were made in the course of establishing the technical baseline; that the government understands what the alternative approaches were and why they were not selected; and that the government agrees with the decisions that have been made.

In the opinion of Gary Kyle, former Air Force contracting executive and president of Persistent Agility, the "government team" is a part of the "acquisition team," as defined in Federal Acquisition Regulation 1.102(c):

> The Acquisition Team consists of all participants in Government acquisition including not only representatives of the technical, supply, and procurement communities but also the customers they serve, and the contractors who provide the products and services.

By "government team," the committee means a team of experts who may be government employees, either civil service or military, or who may be employees of federally funded research and development centers (FFRDCs) (and similar entities such as university affiliated research centers) or independent contractors (but not the prime on the contract).

According to Jon Ogg, former chief engineer at Air Force Materiel Command (AFMC), the composition of the government team may vary from program to program, but the key is that they are all trusted to put the government's interests first in acquisition program decisions. Thus, the committee treated "owning the technical baseline" as a euphemism for having competent and informed government program office personnel and contracted technical support that possess the knowledge, skills, and abilities to effectively execute their programs.

CURRENT CONTEXT OF DEPARTMENT OF DEFENSE ACQUISITION

Beginning in the mid-1990s, there was a series of initiatives within the Department of Defense (DoD), and the Air Force in particular, to reform acquisition processes with the intention to improve efficiency and speed the delivery of weapon systems to the warfighter. The general thrust was to reduce government hands-on oversight of acquisition programs, reduce or remove the government's role as system integrator, and give industry contractors more flexibility and responsibility for program execution. Secretary of Defense William Perry began the process with a shift away from reliance on military specifications and standards in the mid-1990s, prompted by the defense industry campaigning on the platform that onerous military specifications and standards were adding 20-30 percent additional cost to programs but added little to the effectiveness of the final system. In the opinion of one of the participants, Jon Ogg, many of the best practices that served to guide the way that both the government and industry engineers operated were discarded during this time, and now, when there are no comparable commercial specifications or standards available, industry and government engineers are forced to relearn them—through costly trial and error in many cases.

At the same time, an acquisition approach known as total system performance responsibility (TSPR) was initiated to reduce non-value-added government involvement and transfer government tasks to the contractor (with minimal government oversight) in order to gain efficiencies. The idea was that by properly specifying end-item performance requirements, the desired outcome could be attained without imposing particular design approaches or detailed process requirements on contractors. The TSPR approach was used by the Air Force to justify downsizing its acquisition workforce and reducing the number of core functions—in particular, engineering functions—within its program offices by shifting responsibility to industry.

The Principal Deputy Under Secretary of the Air Force for Acquisition's "Lightning Bolts" of the late 1990s and early 2000s, a series of initiatives created to reform the Air Force acquisition and sustainment processes, also worked against sustaining and strengthening key acquisition functions, particularly engineering. System program offices (SPOs) were instructed to reduce personnel to a maximum of 150 personnel, with the notion that the appropriate role of the government program office personnel was to maintain "insight" on contracts rather than "oversight." This led to a major downsizing of the acquisition workforce, with engineering taking the brunt of the reductions, due in part to the belief that if a competent government team, proper insight, correct metrics, adequate funding, and appropriate contract incentives were in place, the contractor team would successfully execute the program.

Many program managers (PMs) of that era viewed government engineering's role as generating the system requirements (Systems Requirements Document or Performance Specification) from the User's Operational Requirements Document. These would then be placed on the development contract, and the government would step back and let industry's engineering workforce deliver. In many quarters, this was interpreted as directing government functional experts to take a "hands-off" approach to managing the technical aspects of the program. In the view of Ogg, the position espoused by senior leadership was that the government should not insert itself into the day-to-day execution of the contract but instead monitor the contractor's performance and reward or penalize the contractor accordingly to ensure that DoD receives the execution performance and product required. In Gary Kyle's experience, the proper definition of "insight" included in the contract was essential to establishing the proper framework for programs that allowed for proper government team engagement and successful TSPR implementation.

Principal Deputy Under Secretary of the Air Force for Acquisition Darleen Druyun testified to Congress on March 17, 1999:

> The USAF is making great strides in decreasing our acquisition workforce. TSPR served to allow the USAF the opportunity to downsize. In 1997, the Air Force re-evaluated its core functions governing systems engineering, reducing the core functions from 20 to 10. Re: under TSPR we have reduced POs from 242 to 20 (ref F117). The new Program Office functions include Requirements, Direction, Budgeting, Contracting, Product Acceptance, and Security. Giving contractors total system performance responsibility, using performance-based requirements, eliminating military specs and standards, and relying on contractor data and metrics are all examples of reforms that have resulted from a change in philosophy from government oversight [involvement] to government insight [monitoring].[1]

Another change to the Air Force engineering workforce came in the mid-2000s with the reorganization of AFMC into wings, groups, squadrons, and flights (W/G/S/F), in the process doing away with SPOs and moving the "ownership" of the program office functional staffs (most notably engineers) to the PMs in charge. In the opinion of one senior Air Force leader who attended the first session, this significantly hampered the functional leads across AFMC's three acquisition centers within the Command—that is, the Aeronautical Systems Center, the Air Armament Center, and the Electronic Systems Center—from performing their role to train, organize, and equip.[2]

Prior to this W/G/S/F reorganization and personnel realignment, AFMC's engineering leads could move engineers across programs as needed, working collaboratively with the program office PMs to provide engineers with the breadth of experience required to function effectively as subject-matter experts or lead/chief engineers. The functional leads could also decide what additional training or education was required for each individual to better prepare him/her for assignments of higher responsibility or to sharpen or augment competencies to effectively function in his/her current role. This organize-and-equip function by the AFMC's three acquisition center engineering staffs came to a halt under the W/G/S/F construct as PMs now owned the positions and could make reassignments independent of approval or input from the home offices. The effect was, in Ogg's view, that movement of personnel by these respective center engineering leads across programs became stalled.

Jon Ogg submits that these "streamlining initiatives" of the mid-1990s through the turn of the century contributed to the atrophy of the technical workforce at AFMC. Furthermore, another participant, Michael Griffin,

[1] Summary of prepared statement of Darleen A. Druyun, Principal Deputy Assistant Secretary of the Air Force for Acquisition and Management, before the Senate Armed Services Committee Readiness and Management Support Subcommittee, Subject: Air Force Acquisition Reform, March 17, 1999.

[2] During this time period, Air Force Materiel Command (AFMC) had three separate acquisition centers: the Aeronautical Systems Center, the Air Armament Center, and the Electronic Systems Center. The AFMC subsequently reorganized again in 2012, and all three acquisition centers were consolidated into a single Air Force Life Cycle Management Center under AFMC. A separate acquisition center for space—called the Space and Missile Center—remained outside of AFMC under the Air Force Space Command.

argued that even PMs came to not understand the technical problems in their programs. Griffin also believes that systems engineering came to be viewed by the Air Force strictly in terms of process. The result, in his estimation, was that the larger issues concerning what the system design should be—and why it should be that way—no longer received the level of attention that had been required in prior generations.

The weakening of the technical workforce was acknowledged by Air Force senior leadership, and in the past 10-15 years, efforts have been under way across the Air Force to "shore up engineering," beginning with Secretary of the Air Force Roche's revitalization of systems engineering in 2002 and continuing with initiatives such as Operational Safety, Suitability, and Effectiveness; Life Cycle Systems Engineering; and the Systems Engineering Assessment Model. In addition, according to Ogg, DoD has restarted the use of selective military specifications and standards to help guide industry and its own engineering workforce with proven practices and processes. David Walker, Deputy Assistant Secretary of the Air Force for Science, Technology, and Engineering, noted that within the past year the Chief of Staff of the Air Force and the Secretary of the Air Force have signed off on a policy[3] to rebuild technical capability, establish standards, and understand core competencies and gaps. However, despite these recent efforts, according to several participants Jorge Gonzales at AFLCMC, who has the responsibility for the career development of Air Force engineers, also has to negotiate with PEOs to move engineers to support programs. Basically, they argued, he does not have complete authority. Bolton noted that there is both a structural concern and a workforce concern that ultimately will need to be addressed. In commenting on this, Bud Boulter, from SAF/AQR, described to the workshop participants a recent CSAF and SECAF directive that attempts to clarify this topic.

A few workshop participants commented after the discussion with Dr. Walker that past policies have resulted in a situation where the Air Force in many cases does not retain enough technical information about its acquisition programs (and, even with outside help from FFRDCs and independent systems engineering and technical assistance contractors, does not have enough qualified people who understand the technical information) to control costs and make smart acquisition choices through the program life cycle. The question is, what is the appropriate balance between the notion of government as system integrator on one extreme and the TSPR approach, in which government's role in the management of the system is minimized, on the other?

As Col Robert Strasser, system program manager, B-2 Division, believes, the Air Force must own the technical baseline for long-term continuity of operations. Over the past two decades, he argued, the Air Force has gone from oversight, to insight, to out-of-sight with its lack of program ownership. Col Strasser also noted that the government should own the technical baseline because contractor experts die, retire, and move to other programs (or refuse to relocate if the prime contractor moves a facility).

REPORT ORGANIZATION

Task 2 in the terms of reference implies that challenges inherent in owning the technical baseline are not just engineering issues but extend to larger aspects of acquisition. In Chapter 2, the observations from several participants are presented about the status of owning the technical baseline in the Air Force in the following categories: programs, leadership and culture, workforce, funding, contracting, and policies and processes. Finally, suggested terms of reference for a potential follow-on study (Task 3 in the terms of reference) are provided in Appendix C.

[3] U.S. Air Force, *Air Force Engineering Enterprise Strategic Plan 2014-2024*, May 2014.

2

Workshop Participant Observations

PROGRAMS

Over a series of three workshop sessions, the participants received briefings from 13 Service programs.[1] Claude Bolton, Jr., executive-in-residence at the Defense Acquisition University, summarized these programs into two categories of technical baseline ownership: (1) the "clean sheet" situation with a new acquisition program, in which the government clearly owns the technical baseline, and (2) the situation where the program is in some process of "regaining" the baseline of an existing capability. The abstracts for these briefings are presented in Boxes 2-1 through 2-5.

The Advanced Pilot Training Family of Systems (next-generation trainer) program known as T-X was briefed at the workshop by Col. Peter Eide. He noted that this program was an example of doing things right. Bolton expressed his opinion that the implementation of the development planning approach, similar to the way things were done back in the Vanguard days, along with the pre-milestone B activities that included technical baseline ownership and activities, are well under way, ensuring that the "right people" appear to be in place along with the right resources (processes, tools, procedures, and techniques)—all supporting his assertion that the T-X program "gets it." The only possible problem, he added, may be an ability to get or keep enough of the right people and funding in the future. Jon Ogg pointed out that the T-X session highlighted that owning the technical baseline (OTB) success is not because of policy, metrics, instructions, or processes. It takes smart people (knowledgeable and competent) making informed decisions to define the acquisition strategy of the program. He noted that they are using the proper tools to evaluate the alternatives while staying engaged with their customer, Air Education and Training Command, to keep them informed of the consequences, that is, development and production cost and schedule impacts as they evolve the T-X requirements.

Claude Bolton made the point that the Family of Advanced Beyond-Line-of-Sight Terminals (FAB-T) program presentation was in contrast to the T-X briefing, in that there was lack of leadership or ownership from its inception, from 2001-2009, which led the program to experience cost overruns of up to $1.2 billion and an 8- to 9-year delay. However, Bolton noted, under the leadership of the chief engineer for the past 3 years, the FAB-T program has been put on track. It appears now that the engineers do "real" engineering work by running analyses and models to verify contractors' deliverables. Richard Roca concluded by adding that it was important to note

[1] See Appendix B for a complete listing of presentations.

BOX 2-1
KC-46

Brig Gen Duke Richardson, Program Executive Officer
Tanker Directorate, Air Force Life Cycle Management Center (AFLCMC/WK)

The KC-46 tanker modernization program, valued at $52 billion, is one of the Air Force's top acquisition priorities. KC-46 aircraft are expected to replace a portion of the Tanker fleet. In order to affordably sustain the KC-46 over its programmed 40-year life span, the Air Force acquisition community ensured adequate information is available to execute the weapon system's acquisition and sustainment strategies. By militarizing a commercial freighter, the Air Force sought to reduce cost and technical risk by leveraging both commercial and military best practices. The Air Force utilized a contracting strategy that protected government interests through requirements development, acquisition of data and data rights, shared risk, and clearly defined technical baselines. Implementing these plans early in the life cycle is providing a technical baseline that balances both near-term fiscal responsibility and long-term sustainment needs.

BOX 2-2
Long-Range Stand Off

Mr. Ken Lockwood, Director of Engineering, Strategic Systems Division (AFLCMC/EBB)

The Long-Range Stand Off (LRSO) cruise missile program, valued at $8.7 billion, is a top Air Force acquisition priority. LRSO is the follow-on to the Air-Launched Cruise Missile (AGM-86B), focused on nuclear deterrent capability. In order to have a sustainable and robust system, the Air Force acquisition community will emphasize a robust design, mature manufacturing processes, and competitively priced production lots. The Air Force will realize a robust design via prescriptive design for reliability and design for manufacturing programs, with emphases on eliminating failure modes as well as increased government participation in the day-to-day activities of primes and suppliers. Beyond a robust design, the contracting effort will focus on clearly defined technical baselines and priced production lots to balance both near-term needs as well as the long-term sustainment of the LRSO cruise missile.

that every successful program presented at the workshop had a program office that had a deep technical insight into the operational environment in which the system would be used.

During the workshop, however, the participants also heard examples of programs that did not have control of their technical baseline. One of the best examples of why owning the technical baseline is important was provided by Col. Robert Strasser, system program manager, B-2 Division. He noted that neither the Air Force nor the prime contractor can reconstruct the inlet on the B-2 due to loss of the technical baseline. Strasser stated that only 18 percent of the organic system program office (SPO) personnel are science and engineering (S&E) professionals, and the prime is actually in the SPO as a systems engineering and technical assistance contractor, which he believes presents organizational conflict of interest and provides no ability for the government to independently challenge the prime's technical risk assessments or cost estimates. And while the program has acknowledged that it understands that the lack of technical expertise within the program office is a fundamental flaw and has started to look for opportunities to hire a lead systems integrator contractor, Strasser stated that personnel policies have blocked any progress on this front.

BOX 2-3
Advanced Pilot Training Family of Systems

Col Peter Eide, Program Manager

The Advanced Pilot Training (APT) Family of Systems (FoS) acquisition program, commonly known as T-X, is one of the Air Force's top acquisition priorities. By 2031, greater than 60 percent of the Combat Air Force fleet will be "5th generation." APT aircraft and associated ground-based training systems, mission planning systems, and courseware will replace the aging 3rd-generation T-38 Talon fleet, thereby providing an essentially higher-quality training experience for our Airman. By leveraging mature designs—11 partner nations currently fly more advanced trainers—the Air Force plans to minimize developmental costs and reduce acquisition risk. Following in the footsteps of pathfinder programs like KC-46, the Air Force plans to maximize insight into and access to design-level data across the program life cycle. Tenets of owning the technical baseline (OTB) will help the Air Force be a more informed customer with the maximum opportunity to compete future modification and sustainment workload. Adopting OTB while also pursuing strategic agility at the outset of the APT program will help to ensure maximum capability at minimum cost.

BOX 2-4
Family of Advanced Beyond-Line-of-Sight Terminals

Dr. Thomas Kobylarz, Chief Engineer

The Family of Advanced Beyond-Line-of-Sight Terminals (FAB-T) presentation described a government chief engineer's interpretation of a system's technical baseline followed by evidence of how it is used to effectively execute a development and production program. The briefing starts with a definition of terms to place the work in context, then a brief background on the FAB-T, a satellite terminal program, which will be used to demonstrate the concepts. Examples specific to the program are given to quantify the system attributes that define the baseline as well as demonstrate the processes used to maintain control of that baseline and support informed programmatic decisions.

The Expeditionary Combat Support System program, Jeff Stanley noted, was designed to replace a multitude of disparate logistics information technology systems. Gary Kyle, however, claimed that the program clearly demonstrated lack of ownership of the technical baseline from inception. Kyle stated, "This program never had a good 'as is' baseline for the 200+ requirements it was supposed to replace." Additionally, he stated, larger problems existed, such as lack of a clear understanding of the requirements; lack of proper tools, processes, techniques, and so on; and, most importantly, lack of the "right people," defined as those who were educated or trained and had the appropriate mentored experience to do the job. Bolton pointed out that lack of the right people included people inside and outside the program management office. "They simply did not know what they didn't know," Ogg observed. Michael Griffin agreed, saying that "there was no technical baseline—no one owned it; it did not exist."

Richard Roca expressed his amazement at the numerous examples of programs where the government funds the contractor for all the work yet does not retain rights to all the intellectual property generated by that funding: "If I understood the presentation, DCGS is an example where the Air Force does not have access to interface information yet is the sole funder of the work. This is incredible if true." Furthermore, Roca noted, this issue is

BOX 2-5
Distributed Common Ground Systems

Mr. Steven Wert, Program Executive Officer
Presentation provided by Col Ray Wier, Chief, C2ISR Division
Battle Management Directorate, Air Force Life Cycle Management Center (AFLCMC/HB)

The Air Force Distributed Common Ground Systems (DCGS) weapon system is a top priority for Air Force acquisition. Air Force DCGS provides a global intelligence, surveillance, and reconnaissance (ISR) capability that processes and exploits U.S. and coalition data. In order to ensure a secure and sustainable system, the Air Force will emphasize rigorous engineering processes, a government-owned integration process, and competitive contracting. By standing up the ISR Systems Engineering Center (ISEC), the Air Force will leverage original equipment manufacturer best practices and initiatives while maintaining government control of integration and testing. This will ensure that all new software and hardware will be vetted through a stringent Air Force-owned testing process before reaching the enterprise. Along with the ISEC, the Air Force will focus on decreasing sole source contracts, driving competition by acquiring data rights, and clearly defining technical baselines. Implementing these plans will balance both near-term needs, as well as the long-term sustainment of the Air Force DCGS Weapon System.

exacerbated by the fact that several of the Air Force's acquisition programs are based on a commercial platform (e.g., the tanker), which clearly raises challenges compared to military programs that do not rely on a commercial system. In this case, Roca believes that defining the ownership of data rights in the early steps of establishing the technical baseline is critical. Ken Lockwood, director of engineering, Strategic Systems Division, Air Force Life Cycle Management Center (AFLCMC)/EBB, added that it is important to recognize that the technical baseline data may exist at many levels. To own it, he added, it is necessary to get to lower-tier suppliers. But, as Bolton pointed out, that leads to an even more fundamental concern for the Air Force—namely, how do we even know the current state of OTB in the Air Force? Because even for the ACAT-1 programs, which are by definition highly vetted and coordinated programs, Bolton thinks that many of the program offices do not have the organic technical capability to provide an independent assessment of the contractor's claims.

Terry Jaggers, former chief engineer of the Air Force and former deputy chief engineer of the Department of Defense (DoD), asserted that the Air Force owns the technical baseline of all programs before contract award, at which time it "conscientiously" keeps or gives up [some or all] ownership to the prime contractor for the life of the program. However, he said, it is his understanding that there is no process to thoughtfully decide the extent of ownership for programs before the prime contract is awarded. While not all programs may require the same level of ownership of the technical baseline, there is no process to capture the trade studies as artifacts by which the Air Force decides how much to own and how much to cede to the prime contractor (or decide the level of some or any joint ownership), nor any evidence of risk management in terms of a reengineering capability or contract language to retrieve ownership, should it be required at some point in the program's future. He thinks it would be important to do this decision process early and conscientiously before contract award, because regaining ownership is a resource-intensive (and perhaps impossible) task. Furthermore, Jaggers added that ceding any ownership to the prime at contract award requires a resource commitment to manage the risks from this decision, should unforeseen exigencies require ownership again (parts obsolescence and integration of subsystems being two prime examples).

As seen in Figure 2-1, from Lt Gen John Thompson, commander of AFLCMC, current programs span the spectrum of ownership, which largely reflects decades of various acquisition reform initiatives where ownership was not a consideration but was rather "baked in" to the process. In other words, he pointed out, a program's position on this spectrum of ownership is what it is as a result of the "acquisition reform" environment at that time, not what it should have been for the long-term life of the program after careful program-by-program delib-

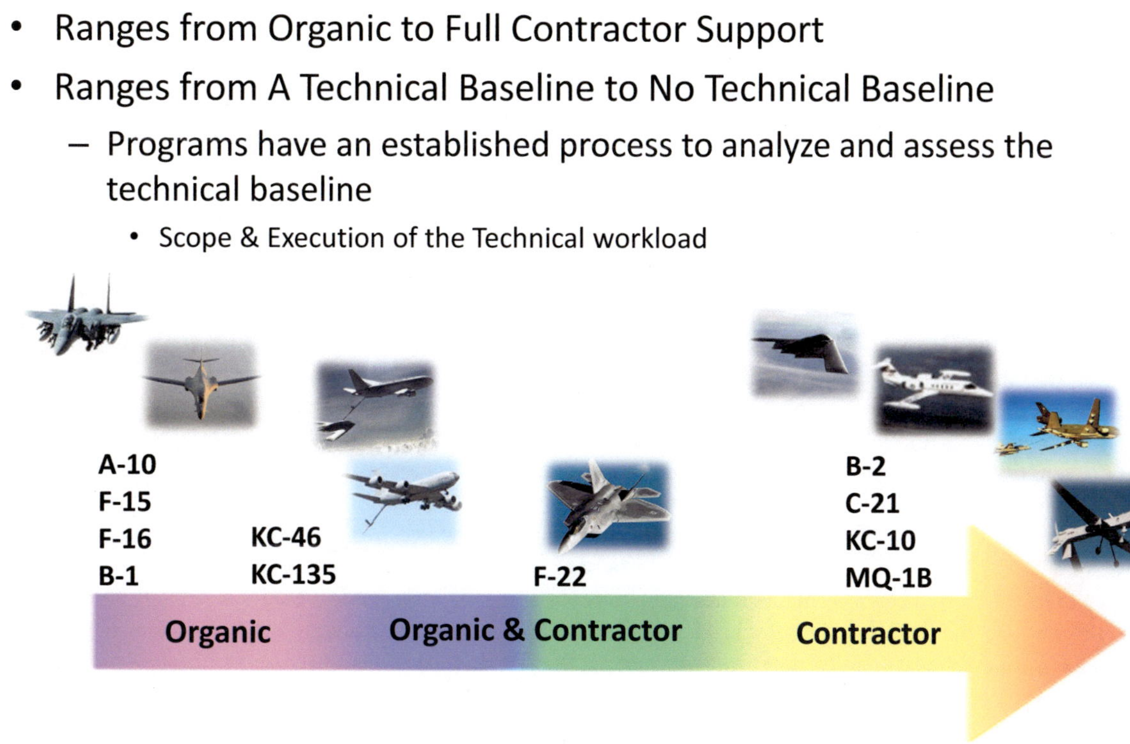

FIGURE 2-1 Defining a program's baseline. SOURCE: Lt Gen John Thompson, Commander, Air Force Life Cycle Management Center. Approved for release by SAF/PAO on December 18, 2014.

eration and decision. Jaggers thinks the Air Force should consider inserting a process of deliberations complete with trades and risk-reduction strategies during the pre-milestone A development planning process and perhaps capture these trades and decisions in the acquisition strategy plan for all future programs prior to contract award. In other words, he stated, ownership of the technical baseline should be a conscientious and deliberate decision process for every program.

Jaggers went on to say that ownership requires a standard and metrics to assess attainment and maintenance of ownership goals for an individual program. Based on the preliminary working definition provided by Col Strasser, he said that the A-10 could serve as one benchmark for a platinum level of ownership, not because the Air Force chose to own it at the beginning, but rather because Fairchild went out of business, and therefore the Air Force had to assume complete ownership at the end. Another participant asserted that there may be other benchmarks both within the Air Force and other services, but the limitations of a short workshop prohibited an in-depth discussion about the desired benchmark end-state vision and appropriate definition for full ownership of the baseline. Once a benchmark standard is created based on A-10 or other exemplar programs, Jaggers suggests that metrics be developed to ensure ownership is maintained and risk appropriately managed over the life cycle if ownership is ceded to the prime contractor. Furthermore, he believes that this should be a consistent status item at all major program reviews.

A recurring topic was the view, expressed by several workshop participants (Terry Jaggers, Trey Obering, Richard Roca), that the most important issue with respect to managing the technical baseline of a weapon system is for the PEO and PM to be able to oversee and manage the baseline with accountability, authority, and responsibility.

LEADERSHIP AND CULTURE

Throughout the workshop sessions, Claude Bolton repeatedly stated that "the Air Force was once the leader in the principles of program management." Jon Ogg and Trey Obering, a former Director and Acquisition Executive for the Missile Defense Agency and currently a Booz Allen Hamilton executive vice president, highlighted that the loss of the technical baseline did not occur overnight; instead, they argued that the erosion of the fundamental ability to manage the technical baseline was the result of the leadership's acquisition policies (e.g., total system performance responsibility, or TSPR) and organizational consolidations that occurred in the mid-1990s through early 2000. According to Obering, to regain ownership of the technical baseline, it will take unified leadership support to establish a policy that technical competence in the program offices is a strategic priority. Obering ended by stating, "absent strong leadership, the Air Force will not get out of its current state."

As noted at the first workshop session by David Walker, initial steps are already under way, in that the Air Force has developed the *Air Force Engineering Enterprise Strategic Plan 2014-2024*.[2] He stated that both the Secretary and the Chief of Staff of the Air Force define, as a high priority, the need to "address engineering enterprise workforce issues, including core competencies, structure, development and assignments."[3] Yet, according to other presentations, a recurring theme among all senior acquisition leaders is that they do not have the authority to rapidly hire and transfer critical engineering talent across programs and PEOs. A request for personnel action to move or hire a civil service engineer takes months to complete. Air Force Headquarters Deputy Chief of Staff for Personnel, HAF/A1, is in control of this, but A1 is not accountable for acquisition program success. Obering noted that reallocation of engineering resources to a higher priority is important to program success but currently has to be executed via the Unit Manpower Document (UMD), which, in his view, is very inflexible.

Lt Gen Thompson provided the workshop his perspectives during the second session (see Box 2-6). Following his presentation, Lt Gen Thompson stated that only the Secretary and Chief of Staff of the Air Force can change the current policy to direct that UMD management should reside at the AFLCMC level, which would provide the flexibility for his functional leads to control the movement, training, and promotions of their technical workforce. Obering noted that while the release of *Air Force Engineering Enterprise Strategic Plan 2014-2024* represents a major and positive step forward, it remains to be seen how fully it will be implemented and what resources it receives.

Richard Roca agreed but stressed that to be effective, OTB must be a top-down cultural mindset originating at the highest levels of Air Force leadership and promulgated throughout the military ranks and the civilian staff. Sue Payton, president of SCI Aerospace, Inc., reminded the workshop that the presentations of RADM Dave Johnson, PEO Submarines, and VADM Terry Benedict, Director Strategic Systems Programs, were particularly powerful illustrations of the positive results that can be achieved when (1) there is a strategic long-term commitment from leadership to maintaining ownership of the technical baseline and (2) acquisition leaders are vested with the authority to retain, train, and plan personnel succession and are accountable and responsible for program success. This mindset, as expressed by Griffin, should include a core belief that the U.S. government, and particularly the Air Force, must fully embrace the need to control its destiny in regard to the acquisition, use, and support of its weapon systems and the organizational and logistical elements that support them. He continued that this mindset, if properly embraced, will necessitate changes in contracting, staffing, training, and promotion philosophies both within the military ranks and within AFLCMC support organizations, including the Air Force Research Laboratory, federally funded research and development centers (FFRDCs), university affiliated research centers, and professional service contractors.

Jon Ogg and Sue Payton believe the Air Force seems to be conflicted in its position on the value of a strong and effective engineering and technical workforce, for it is not clear that there is a unified position across Air Force program offices as to the need, role, or value of an organic engineering workforce. For example, as noted above by Lt Gen Thompson's comments about the reorganization of Air Force Materiel Command (AFMC), the engineering functional home office (AFLCMC/EN), charged with training, organizing, and equipping AFLCMC's engineering and technical workforce, does not retain the necessary control (UMD ownership) over its workforce

[2] U.S. Air Force, *Air Force Engineering Enterprise Strategic Plan 2014-2024,* May 2014.

[3] Ibid.

BOX 2-6
AFLCMC Initiatives and Perspectives

Lt Gen John F. Thompson, Commander, Air Force Life Cycle Management Center (AFLCMC)

For AFLCMC, our execution and functional portfolios span the spectrum from full organic support to full contractor support and everything in between. On the surface, the technical baseline is composed of data and information that provide the program office knowledge to establish trade-offs and verify, change, accept, and sustain functional capabilities, design characteristics, and quantified performance parameters at the chosen level of the system hierarchy, but "ownership" of it is much more. It ensures programs have established processes to analyze and assess weapon systems to enable informed decision making. Key points that influence the technical baseline in the Air Force are policy; types of funding; keeping appropriate knowledge, skills, and abilities; adequately resourcing programs and processes with manpower; and providing management and engineering tools on demand.

to allow him to make the moves or mandate and provide the training of AFLCMC engineers assigned to programs to ensure they are strong technical leaders that are seen by the PM and industry counterparts as "Heisman-level" players on their acquisition team.

However, Griffin observed that in spite of this absence of a clear Air Force mandate, there is a palpable sense of "ownership" by Air Force PEOs/PMs for their programs, which is clearly present in some, possibly even a majority, of workshop presentations. This sense of ownership is evidenced by some "should be" standard for all, Griffin asserted. Importantly, and very obviously, those PEOs/PMs who embraced such ownership were doing so not because of Air Force leadership policy, but in many cases almost despite the best efforts of Air Force and DoD seniors to militate against their efforts. Dedicated PEOs/PMs, Griffin noted, were clearly seeking "workarounds," authorized or otherwise, to compensate for lowest price technically acceptable (LPTA) or small business set-aside procurement limitations, civil service hiring practices and limitations, or uniformed service staffing and promotion policies. He said that DoD, and particularly Air Force, policies and procedures should not be such as to impede PEOs/PMs from doing obviously correct things to enable mission success. However, during his presentation, Stan Soloway noted that there is a major disconnect between the messages from senior DoD leaders and the lower levels of management. One participant observed that it is no wonder the PEOs/PMs are confused, as it had been reported that Frank Kendall, the Under Secretary of Defense for Acquisition, Technology, and Logistics, had stated publicly that LPTA is not the way to contract for complex systems, while at the same time his subordinates were imposing LPTA as the solution for all things.

As Terry Jaggers previously noted, it appears that a "gold standard" for owning the technical baseline might be something like the A-10, and while the Air Force is forced to own it because the contractor is gone, it still serves as a much-needed standard or benchmark. Jaggers went on to say that whatever the gold standard, institutionalizing the ownership process and degrees of ownership is a leadership challenge. It will require a long-term cultural change before it is embraced by not only the acquisition cadre but the support functionals, and whatever standard is created should be institutionalized to outlast the current service acquisition executive (SAE). Jaggers voiced his greatest fear that the high standard set today by the current SAE will change in a few years and with each subsequent SAE. The Air Force, Jaggers believes, should try to set a technical baseline standard so that a foot of ownership equals 12 inches, not "the current king's shoe size," and this standard should be inculcated in all leadership functional training.

A recurring topic was the opinion, expressed by at least two participants (Michael Griffin, Trey Obering), that it is essential that senior Air Force leadership make it clear to all functional leaders supporting acquisition that the Air Force highly values technically trained and competent acquisition and engineering personnel.

WORKFORCE

Mike Griffin stated that a point that resonated during the course of the workshop was that "managing an engineering development program requires engineering talent because requirements cannot be written so carefully as to ensure program success and contractor performance." However, as noted by Sue Payton, it is disheartening to find that professionally trained engineers comprise only 20-25 percent of Air Force engineering development programs. It is also of great concern, she noted, that the role of government engineers in the most important programs, such as the KC-46, is to witness, review, observe, and take passive action. In many cases, Air Force engineers are not really doing any engineering on the program; they are merely watching as others do. As described by Griffin, "They are art critics, not artists." Payton then asked, "Who has the responsibility for growing the competence of the Air Force engineering workforce?" She stated that she has observed that the technical competence of the Air Force acquisition workforce has deteriorated dramatically because external organizations such as HAF/A1, SAF/AA, and SAF/FM have usurped the authority from SAF/AQ to manage acquisition personnel assignments, resources, and promotions. "Who are the custodians of the Air Force engineering workforce and culture?" questioned Payton. During the discussion of Aegis—the Navy's weapon system that uses powerful computer and radar technology to track and guide weapons to destroy enemy targets—Kate Paige and Don Mitchell both described the Navy's view on what OTB meant for its programs. RADM Paige elaborated that it means the government is accountable for the entire systems engineering process and its outcomes (see Box 2-7). Both Mitchell and Paige agreed that the Applied Physics Laboratory at John Hopkins University had played a key part in the government's role as systems integrator, which, as one participant noted, was a very different management model from the Air Force uses. Another distinct difference between the Navy and Air Force, according to the two, was in the length of time an officer spends throughout his/her career working in a specific program. Paige commented that she had been associated with Aegis since 1976.

During this session of the workshop, Bolton stated that one cannot discuss OTB without positing that there is a cadre of experts on the government team who are capable of understanding the data and using it to make informed decisions. He continued by asserting that in the past the Air Force had, broadly speaking, access to the technical talent (uniformed and civil service staff, FFRDC, or professional service contractors) that is necessary for owning the technical baseline. However, as Griffin observed, in many of the presentations over the course of three sessions, the Air Force did not possess the capacity or tools to provide an independent assessment of prime contractor claims, judgments, or performance. In many cases offered at the workshop, the prime contractor was the primary or even the only source of relevant information. Obering believes that it will require the better part of a generation to remedy this situation, even with the full support of top Air Force leaders. One possibility might be to rely on the use of FFRDCs as a transition mechanism for the program offices as they regrow their organic capability. But as Ogg pointed out, the penetration of the FFRDCs is limited because they primarily support East and West Coast installations. For installations in lower-cost-of-living areas where DoD pay is comparable to local industry, he added that it is probably better to use the civil service hiring process to bring aboard strong technical journeymen and grow the workforce organically through formal education, hands-on experience, and rotational assignments in the Air Force Research Laboratory (AFRL). However, this approach is also problematic due to the lethargic civil service hiring process, where it can take upwards of 6 months to hire an individual and little or no funding is available to allow the new employee to advance his/her academic education or participate in other forms of training, such as attendance at technical conferences, exchange with industry, or technical short courses. Additionally, Ogg noted that when a downturn inevitably hits, the recently hired junior staff are the first to be "surplussed" because DoD uses the last-in-first-out method that surpluses the most recently hired. This sends major repercussions through the universities that had encouraged their graduates to seek employment at the local Air Force installation. Jaggers observed that there may be a potential to leverage some aspects of the various workforce demonstration programs to alleviate several of the aforementioned issues, most notably the laboratory demonstration (Lab Demo) program or acquisition demonstration program authorities for things like expedited hiring. Jaggers added that the Air Force should also look at how the Navy in particular uses Lab Demo authorities to further the interests of the broader Navy programs, not just Navy science and technology workers. This review could be completed in the proposed terms of reference for follow-on study when reviewing other Services' challenges.

BOX 2-7
Owning the Technical Baseline in the USAF

RADM (Ret) Kate Paige, USN, w/ Mr. Don Mitchell (JHU/APL)

RADM Paige and Mr. Mitchell discussed what "owning the technical baseline" means in the Navy's Aegis program and MDA's Aegis BMD program and how it is accomplished. They offered their historical insight as well as relaying the views of the current programs' leadership. To support the discussion, they provided handouts that addressed a notion for organizing the major players in Defense acquisition, along with comments provided by the current program offices.

Ogg believes the Air Force leadership must realize that it is going to be an uphill battle to capture, retain, and train organic technical talent within the acquisition workforce. AFRL does a much better job of attracting, continuously training, and retaining technical personnel, some of who provide advice and counsel to acquisition program offices. Ogg went on to point out that as for engineering support contractors—often hired through advisory and assistance contracts (A&AS)—the contracting community has driven to LPTA suppliers. Once again, this makes it difficult to bring aboard top-notch technical talent because they find other opportunities working for industry rather than signing on with local contractors who pay minimum wage in order to secure Air Force Service contracts. Ogg also noted that support contractors cannot represent the Air Force in key leadership positions such as chief or lead engineer. So a conundrum exists: (1) FFRDCs that have limited geographic distribution, are expensive, and are capped by Congress; (2) the challenges of attracting, hiring, training, and retaining civil service talent; and (3) low-bid contractors who provide marginally acceptable technical talent. The result, Ogg concluded, is the current state of disillusionment and dissatisfaction with the acquisition technical workforce—"a perfect storm in the making!"

To fix the situation and have the capacity to own the technical baseline for Air Force programs, a key element should include changing the culture of how the Air Force values its engineering workforce, Ogg believes. As Kyle and Jaggers noted, the technical workforce includes military, civil service, and contractor S&E professionals. They stated that the Air Force needs a unified, holistic program-staffing policy that includes defining who is responsible for ensuring that for all Air Force programs the technical baseline is known and maintained for the life of the program. Unfortunately, Obering pointed out, the situation is especially dire for the uniformed members of the S&E workforce. As it stands now, there is no career path in the military clearly outlined for engineers that includes flag-level officers. He went on to point out that on the military side no career path exists beyond the rank of major for the engineering field—that is, there is not an "engineering duty officer" approach to three-star rank such as the U.S. Navy employs. Further complicating successful career upward mobility, Payton observed, is that members of the S&E workforce seeking to remain in the military must move out of the Officer Air Force Specialty Codes 61 and 62 to have promotion opportunities. Ogg suggested that perhaps the Air Force should create a bridge program that would allow for those in the military desiring to remain in engineering to move from military to civil service after putting in their minimum military tour. These individuals are high-value assets due to their military training, familiarity, and identity with the ultimate customer and their core values—integrity, service before self, and excellence—in all they do. Roca added that these individuals also bring experience in the relevant operational environment to the program office, which appeared to be a hallmark of the successful programs presented during the workshop.

Throughout the workshop, there were several examples of successful programs where the PM had extensive operational experience with the system. Col Janet Grondin discussed her program at Space Missile Command (SMC) and highlighted the importance of mentoring throughout her career (see Box 2-8 for additional details). Griffin and several of the other workshop participants agreed that SMC is doing a good job and that there clearly are good examples of programs that are running well at AFMC. Obering concluded by stating that SMC seems to prize science and engineering expertise and experience more than other Air Force product centers.

BOX 2-8
Launch Test Range System

Col Janet Grondin, Chief, Range and Network Division (SMC/RN), Air Force Space and Missile Systems Center

Space and Missile Systems Center Range and Network Division (SMC/RN) is the program office responsible for the Modernization and Sustainment of the United States' premier Launch and Test Range System (LTRS). LTRS consists of the Eastern Range (ER), operated by the 45th Space Wing (SW), and Western Range (WR), operated by the 30th SW. LTRS ranges support Title 10 U.S.C. 2273 for Assured Access to Space for National Security Payloads and are Major Range and Test Facility Bases (MRTFB) providing test range infrastructure for spacelift, ballistic missile, guided weapon, and aeronautical missions and systems. LTRS is responsible for the safe tracking and positive control of vehicles flying over thousands of square miles of broad ocean areas in the Pacific and Atlantic. Capture and government ownership/management of the full technical baseline began in 2009. LTRS benefits gained include the award of a fixed price range operations, maintenance and sustainment contract saving $700 million over the 10-year contract, award system modification contracts to best athlete, encouraging competition, improved Reliability, Maintainability, Availability and Dependability (RMAD) data collection and decisions for future system modifications.

Gary Kyle observed that while experience is essential for growing a college graduate into an engineer capable of making informed recommendations or decisions on the part of the program, within the Air Force no such mechanism seems to exist. Hands-on training or experience has become the exception for many entering the Air Force acquisition workforce. Ogg noted that the "Education with Industry" avenue for obtaining hands-on experience working real-world design and test challenges for those in the functional disciplines, especially engineers, has all but disappeared due to cuts to the training budgets. In addition, rotational assignments to AFRL to gain or refresh technological skills are problematic due to different personnel and pay systems and thus are an exception rather than the desired norm. Obering believes that the "Education with Industry" programs for engineers should be re-established, possibly using cofunding with industry. Engineers need to have periodic hands-on experience at the AFMC sustainment center and/or test centers to keep their skills fresh and applicable. Payton wholeheartedly agreed and added that without engineering "know how" PEOs/PMs do not have the intellectual capital to be responsible for the program technical baseline. She said that for technical specialists supporting a PM to be useful, they have to be technically competent, and this requires training at the beginning of their career and continuous technical refresh through education and practice throughout their career. Payton noted that for practice to be relevant, it has to involve working at levels of detail far beyond that achieved when one is observing and commenting on the work of others.

Unfortunately, as Payton reminded the participants, even for the existing technical workforce, there does not appear to be a single organization or SAE that has the ultimate authority and responsibility to ensure that qualified technical resources are available to support the program offices. As a result, SAF/AQ is dependent on critical engineering talent for program success but does not have the authority to move engineers to fill critical program needs. This authority, stated Payton, is vested with the Air Force Personnel Center. The Air Force appears to have lost the flexibility to move engineers where they are needed due to the UMD paperwork approval requirement, observed Kyle. According to Ogg, if the Air Force leadership values technical competence and capability for acquisition success, they have it within their means to invest in the future workforce by providing enablers to acquiring, training, and retaining high-caliber technical talent. He believes that the program office engineering leads should articulate the program's needs, and the functional home office leadership should work to supply the required capability through a combination of organic and contracted resources, with the program office supplying the funding for the contracted personnel. Ogg believes that the functional home offices must regain control of their technical workforce to include resource allocations (assignments), training, and promotions. He said that the UMD should reside with the home offices in order for maximum flexibility in managing the workforce to avoid the onerous paperwork required for personnel reassignments across AFLCMC. Finally, Ogg commented, the functional home offices across AFMC, especially AFLCMC, should be held accountable for sourcing the program offices with the resources required to effectively execute their mission, independent of whether the source is organic or contracted.

Sue Payton agreed with Ogg's suggestions and added that it is fundamental to acquisition program success and to intellectually owning the technical baseline to have the autonomy to manage the very engineering workforce needed for the job. She noted that while the acquisition chain is waiting 6 months or more for civil service personnel to be hired and reassigned, it is losing crucially needed engineering talent to other components within the DoD or industry who can direct hire, despite appeals of patriotism, stability, and education, among other factors, inherent to government work. The responsibility to grow the technical competency of the engineering workforce is diffused, underresourced, and inadequately supported with engineering tools. Jaggers observed that redressing this situation could take a decade, and because the most vulnerable budget items in program offices are systems engineering and training, it will take a renewed commitment from Air Force leadership to rectify this situation. Moreover, making this challenge even more formidable, Obering noted that as a result of past policies, "we've lost a generation of leaders capable of mentorship that could grow back the technical workforce within the acquisition community." Ultimately, the issues highlighted in the discussions related to gaining back the technical baseline can be summarized by a comment made by Col Michael Meyer, deputy director for engineering, Office of the Deputy Assistant Secretary of the Air Force (Science, Technology, and Engineering): "There is a collision between the top-down mandate to own the technical baseline and the bottom-up loss of people and resources."

A recurring topic was the view, articulated by three participants (Terry Jaggers, Jon Ogg, Sue Payton), that continuity, longevity, and mentoring in the engineering and technical fields, including a succession pipeline, are crucial for success of a program. They added that the Air Force needs to implement a formal, robust, and credible training and mentoring program to (1) transfer knowledge to upcoming acquisition professionals and (2) develop demonstrable business acumen in the acquisition workforce.

FUNDING

At various points during the three workshop sessions, Terry Jaggers argued that it may be cost prohibitive for some existing programs to regain ownership of the technical baseline at this point. Jaggers noted that while an aforementioned process should be created to conscientiously decide ownership strategy before contract award, the Air Force should consider a business case on existing programs before directing programs to expend resources to regain ownership. Foremost, he suggested the following should be considered: the type of program (Is it a core military aerospace power program requiring ownership like a fighter or bomber, or a commercial commodity program like a business system?) and the cost in terms of manpower and contract implications to regain it. Jaggers summarized his suggestion by asserting that at present, it seems all legacy programs want to regain ownership, and it is not completely clear that all programs require the expenditure of resources to do so. Regardless of how the Air Force decides to proceed with defining the process for determining how to own the technical baseline on existing programs, he felt that at a minimum, it should realign funding for development of the acquisition workforce. Payton then recommended using research, development, test, and evaluation (RDT&E) funding (otherwise known as Air Force budget account code "3600" funds) instead of operations and maintenance (O&M) dollars (otherwise known as Air Force budget account code "3400" funds) to fund functional training of engineers, which would eliminate the current situation where acquisition training competes against readiness challenges.[4] Kyle concurred that there is a need to explore more use of 3600 (RDT&E) funds versus 3400 (O&M) funds, which has the additional benefit in that 3600 funds allow for more flexibility in hiring contractor technical support. However, Obering asserted

[4] For definitions of 3400 and 3600 appropriations codes, see the Defense Acquisition University's website ACQuipedia, "Research Development Test & Evaluation (RDT&E) Fund," for 3600 codes and "Operations & Maintenance Funds," for 3400 codes, https://dap.dau.mil/acquipedia, accessed April 21, 2015.

that this suggestion would "undo" recent decisions by the Air Force financial community. He added that when Air Force leadership moved funding for AFMC engineering support function from 3600 funds to 3400 funds to allow for a "bill-paying" mechanism, it impacted the ability of AFMC to properly educate and stabilize the technical workforce. In spite of recent decisions that are counter to this suggestion, asserts Ogg, serious consideration needs to be given to converting the organic workforce over to 3600 funding, similar to the way the engineers and scientists in AFRL and test centers are funded. He said this is also true of the Navy's engineering workforce. This would allow more flexibility in the hiring and training of the organic engineering workforce. He argued that the limitations of O&M funds create barriers to the hiring and training of organic resources, whereas the use of 3600 funding would allow the necessary engineering resources (most immediately civil service engineers) to be secured, trained, and employed in the support of program office needs.

The presentations from SAF/AQR and AFMC/EN both described a series of initiatives developed to rebuild the competency and capability of the engineering and technical workforce, yet little funding has been identified by SAF/FM, SAF/AQR, or AFMC to support the work required to bring these to fruition. Bolton and Ogg agreed that if a strong and capable engineering workforce is important to the Air Force, mission funding should be provided to help rebuild the workforce, which has atrophied over the past 10-15 years. This includes hiring top-rung engineers from academia or industry, funding advanced education and training, and having mentors and coaches guide careers through increasingly more important roles and responsibilities and ensuring that engineers gain the necessary core competencies to fulfill the needs of PEOs and PMs charged with executing programs.

While eligibility for tuition "assistance" at local colleges and universities may be adequate, Payton argued, in many instances it is not sufficient to maintain and acquire the level of technical engineering skills urgently needed, as stated by the Secretary and Chief in their recently released *Air Force Engineering Enterprise Strategic Plan 2014-2024*, "to continue to lead the world in the development of those cutting-edge weapon systems vital to the security of our nation and its allies."[5] Payton believes Air Force senior leadership must step forward with adequate funding to support their stated goals and priorities. Additionally, asserted Payton, there are serious gaps in certain core competencies of the engineering workforce such as radar engineers. Without sufficient continuing education funds to allow the Air Force to fully fund, in residence, advanced engineering degrees from highly respected graduate schools in areas critical to future Air Force requirements, Payton asserted that talented engineers will either leave the Air Force or lose their engineering edge. In the absence of Air Force funding, Kyle suggested that perhaps SAF/AQR or AFMC/EN should explore executable approaches to using the Under Secretary of Defense for Acquisition, Technology, and Logistics funding to make focused improvements in educating and training the acquisition workforce. Or, in an example offered by Roca, the Navy has used two concepts successfully: engineering development offices and working capital funding as it relates to the engineering support laboratories used by the program offices. Specifically, Roca noted that the Navy has been able to maintain and grow competencies in various technical areas supporting acquisition ranging from ship mechanical systems, such as pumps, to advanced weapons systems, such as missile guidance systems, by maintaining engineering-based organizations, both within the government and in allied organizations such as university affiliated research centers. These skills are used across a number of programs. Roca went on to say that technical organizations such as the Naval Surface Warfare Center assign individuals to programs based on program need and charge these programs for the use of the staff and equipment through the use of "working capital funds." Fluctuations in need are the responsibility of the warfare center to manage, and they have been able to do so effectively. PEOs are not charged with maintaining the number and competencies of the technical staff and expect the warfare centers (and/or UARCs) to do so. In Roca's view, this system works as well as it does because the typical Navy PEOs are staffed with people who are technically competent themselves, know what support they should expect, and demand it.

A recurring topic was the view, expressed several participants (Claude Bolton, Jr., Jon Ogg, Sue Payton), that the limitations of O&M funding and the inability to use RDT&E funding for hiring, retaining, and training the technical acquisition workforce create barriers to success. As a consequence, they noted, this lack of adequate and

[5] U.S. Air Force, *Air Force Engineering Enterprise Strategic Plan 2014-2024,* May 2014.

timely funding limits the ability of acquisition-center functional leads from shaping the workforce to meet the demands for knowledgeable and experienced technical talent.

CONTRACTING

Terry Jaggers argued that the Air Force and/or DoD should investigate tools and policy to better integrate technical risk assessment by engineering and contracting. From the workshop presentations, he noted that the engineering and contracting functionals work separately with different agendas not only at the beginning solicitation stage, but during the proposal evaluation stage as well. For instance, from the presentation on the LPTA contract evaluations, it appears there is a technical acceptance evaluation done first by the engineering team, then a lowest-price evaluation performed by the contracting or cost team. To fully integrate the technical and contracting team, Jaggers asserted that policy should promote a concurrent evaluation of technical merits simultaneously with cost, and tools should be developed to analyze this joint technical and cost evaluation with some sort of integrated stratification of "realism" rather than pass/fail for technical, followed by lowest-cost selection. The analogy of evaluating proposals for both technical quality and cost concurrently would be similar to the schedule risk assessment process used by the Office of the Secretary of Defense programs. In addition, Jaggers argued that the Air Force could consider a separate "white collar engineering" contract to help struggling programs regain ownership during organic recapitalization at the earliest opportunity. Currently, AFMC and AFLCMC combine all of their technical support under an "Assistance and Advisory Services" label that includes everything from administrative support to professional engineers. While LPTA may be appropriate for administrative support, he stated that it is clearly not going to allow the Air Force to obtain access to the high-end engineering talent it needs to regain OTB. Therefore, Jaggers believes that a separate contract vehicle should be considered for this talent.

There is a clear "disconnect," noted Griffin, between those at the higher levels of acquisition and contracting and individual PEOs/PMs. The former seem to believe that Air Force top-level policies are broadly and fully understood, that such policies have sufficient discretion—"wiggle room"—to allow intelligent program management decisions to be made, and that waivers for specific concerns are routine and easily obtained. The latter group, who are on the receiving end of top-level policies, he argued, clearly believe to the contrary. Such policies, Griffin stated, are seen by PEOs/PMs to be narrowly defined, rigid, restrictive, and essentially non-waiverable. Another major concern voiced by Kyle is that the LPTA contracts do not have the correct skills and experience criteria that allow contractors to provide the needed high-quality and specific technical resources. It appeared that some of the existing LPTA contracts were written to accommodate a larger group of contract awardees, including small businesses, and save money through artificially suppressed labor rates versus ensuring the ability to quickly and efficiently hire proficient, competent technical contractor resources.

Trey Obering reflected on the current contracting situation in which AFMC and AFLCMC mandated use of a specific contract vehicle, ACCESS, for A&AS support. This vehicle is an LPTA Small Business Set Aside contract with the average labor rate less than $45/hour. As Obering observed while driving to the base during a workshop held in Dayton, Ohio, last year, there were signs for gutter cleaner personnel in the area offering their services for $63/hour! He said, this does not bode well for the Air Force to obtain access to the type of talent it needs to regain OTB. AFMC recently started acquiring technical support services using the newly awarded Engineering, Professional, and Administrative Support Services (EPASS)/OASIS contract. The contract, he said, is supposed to be a "best value" approach instead of an LPTA contract vehicle. However, to date, the task orders awarded under EPASS/OASIS have been the same "dive to the bottom" seen under the previous ACCESS contracts. Bolton concluded the discussion by adding his concerns about using this mechanism to assist with maintaining the technical baseline: "Is it even possible to get high-quality and talented engineers on an LPTA contract vehicle?"

Other contracting mechanisms used in the past, such as the TSPR, do not work, asserted Griffin, as demonstrated by the problems that the Joint Strike Program Office has been forced to rectify to get the program back on track. Kyle countered with the perspective that TSPR worked exactly as planned, with appropriate government engagement through an aggressive insight approach, on the Evolved Expendable Launch Vehicle program. Regardless of what contracting mechanisms are imposed by the contracting officer, Kyle argued, the PM and engineering staff must raise issues regarding ineffective contract vehicles to the highest level necessary for valid action and redefining of the requirements versus accepting an untenable or dysfunctional situation. Obering reminded the participants that several of the program offices had expressed frustration that the contracting officer is not subject to any accountability for program success or failure as a result of the suitability of the contract strategy—just the terms and conditions of the contract. Therefore, there is a disconnect in incentives between the PM who needs the technical support and the contracting officer who is not judged on obtaining the technical support needed for program success but rather on whether the contract award is protested or not. For example, Obering continued, LPTA was not intended to be mandatory, but it has evolved that way; SPOs and contracting officers should exercise their own judgment in determining whether an offered price is too low to accomplish the work—not just accept the contractor's low bid as gospel.

Sue Payton echoed Obering's observation that the fact that contracting officers are not held accountable for program success—yet have the authority to constrain the PM from hiring the best engineering talent to support the program—is an issue that the Air Force must address to own the technical baseline. The goal, to own the technical baseline, is clearly achievable in the next decade, but it could be inconsequential to program success if the mission is ultimately compromised due to a contracting strategy enforced by a contracting officer intent on cost control as the paramount metric. After presentations by senior contracting officers and PMs, Payton said it is apparent that the contracting officers do not view themselves as having significant responsibility for successful program execution, only for a successful contract award and protest avoidance. Contracting officer mandates that dictate that a specific contract vehicle be used (i.e., LPTA for acquiring S&E support) can and have precluded the PM from hiring highly skilled engineering talent. Payton concluded her comments by highlighting that for a program office to be effective, the contract strategy must accommodate risk assessments, management, and control.

Both Griffin and Roca noted that the relationship between contracting and program offices in the Navy is significantly different from what the participants heard from Air Force program offices. In the Navy programs presented at the workshop, contracting officers in a properly run program report to the PM. Griffin concurred, "It appears that in the current era that Air Force contracting officers wield inordinate power over a program above and beyond the power of the Air Force PMs." Roca cautioned that contracting officers do have a functional responsibility to exercise contracting processes. They also have a responsibility to ensure legal compliance. He noted that the Air Force does not seem to have distinguished between these two responsibilities for a PM. He said that it is important for a contracting officer to be able to both follow the PM's direction in the day-to-day execution of the program and to also throw a flag if the PM does something that violates procurement law. This is a similar responsibility that every functional support entity in a program serves, but the PM should lead the program in controlling cost, schedule, and performance using appropriate contract types and industry incentives and awards. Kyle suggested that if the contracting officer and PM accept the Federal Acquisition Regulation "acquisition team" approach, then the contracting officer also has an inherent responsibility to ensure program success.

A recurring topic was the view, articulated by two participants (Terry Jaggers, Gary Kyle), that it is essential that contracts reflect the proper level of government technical and business engagement to include oversight, insight, data rights, and intellectual property consistent with the program's life cycle acquisition strategy. Furthermore, they added, this should start at the earliest phases of the program and include a maintenance or sustainment approach for owning the technical baseline over the life cycle of the program.

POLICIES AND PROCESSES

The discussion turned toward what immediate changes to existing internal Air Force policy and processes would help support the OTB objective. During the first workshop session, Jeff Stanley had discussed the Air Force Acquisition Incident Review Board process that had been modeled after the safety review boards. He stated that a central issue in the programs that had been reviewed so far was a lack of good engineering fundamentals. The lack of engineering competence could easily be identified early on, according to Ogg, if major reviews with Air Force and Office of the Secretary of Defense staff and leadership in the Pentagon, at a minimum, included the PM and the program office functional leads for the areas of focus in the review. During the review, Ogg continued, there should be an emphasis on discussing the technical baseline at a level sufficiently detailed to demonstrate the PM's technical competence. In addition, Ogg pointed out that in the programs deemed successful, the chief engineer accompanied the PM to all major presentations, and the chief engineer knew and could articulate the program technical issues. In addition, Kyle thought it would be good for the Air Force to go back to enforcing engineering standards while accepting commercial standards whenever they make sense based upon the acquisition strategy.

Related to tools that could be adopted to gain back the technical baseline, Payton reminded everyone that throughout the workshop discussions with senior acquisition leaders and functional engineering leads, it was reiterated that there is a lack of critically needed engineering tools and modeling software. There is a crucial need to upgrade from a few seat licenses to site licenses for the tools and software that do exist, so that engineers can have ready access to standard engineering tools and modeling software that can be used across programs. Several of the workshop participants were surprised to learn that some financial management officers do not allow engineering tools and software bought by one program to be used across programs. Here again is an example, they pointed out, of an organization (SAF/FM) not accountable for acquisition success with the authority to impact acquisition program success. A variety of standard engineering tools and software should be bought at the enterprise level and then, Payton recommended, pushed out to all the program offices.

Sue Payton expanded on this a bit more by suggesting that while the Air Worthiness engineering process is well defined, there is not a robust, standard collection of engineering processes that other programs follow. As a result, she said, programs adopt the processes, tools, and modeling software of the prime contractors that are supporting the program office. Payton stated that if the Air Force is intent on owning the technical baseline of programs, then the Air Force must independently adopt standard engineering process, engineering tools, and modeling software, used Air Force-wide, to independently assess a program's technical performance.

Terry Jaggers elaborated on this theme by suggesting that perhaps the Air Force should explore a "PEO certification" process similar to that of the Navy's Strategic Missiles or Submarines PEO. He noted that for those critically important military-unique programs near and dear to the Navy's heart, like strategic missiles or submarines, where ownership of the technical baseline is clearly a priority, the PEO has a "personal" certification responsibility to guarantee ownership of the system before it is deployed. This certification is vested in the PEO only, focusing accountability, responsibility, and authority. Jaggers said that it is unclear from the presentations if the Air Force has any similar process on those programs that reflect core military competencies of the Air Force where enduring ownership is valued (such as fighters, bombers, munitions, etc.). As both VADM Terry Benedict and RADM Dave Johnson discussed in their presentations on the Ohio class replacement program (see Box 2-9), the Navy has four personnel accountable for nuclear programs and has vested this accountability into a personnel certification process. In contrast, Jaggers noted that the Air Force often has numerous people responsible for a program, so it can be difficult to codify responsibility for the program's success or failure of maintaining ownership of the technical baseline. On programs that require OTB, Jaggers noted that a PEO certification process could be considered either prior to prime contract award (reflecting responsibility and accountability to manage to the prescribed technical baseline) or before deployment (reflecting responsibility and accountability that the program was developed to ensure the prescribed technical baseline was instilled).

At the conclusion of the third workshop session, several participants reiterated that in contrast to the authority that the Navy's program managers exercise, the rules under which AFLCMC operates impede its ability to acquire, grow, retain, and make available the right staff for programs. These participants believe that for the AFLCMC resourcing system to work efficiently, the commander, who is responsible for organizing, training, and equipping the program offices resident within AFLCMC, needs the flexibility to expeditiously hire, move, and train the workforce.

FINAL THOUGHTS

Several participants noted that ownership of the technical baseline is an Air Force-wide effort involving several functional offices—the government team—not just the acquisition community or program office alone. They believe that the plan and end-state required should be resourced and managed properly by all functionals in the Air Force. Other workshop participants noted that it is not clear that there is unity of effort across the Air Force right now. Within program offices, some believed ownership of the technical baseline should be addressed holistically to include engineering, the PM, A&AS contractors, and FFRDC support to meet ownership goals. Outside

BOX 2-9
Ohio Replacement (OR) Program

RADM Dave Johnson

The Navy's Ohio Replacement (OR) will replace the current fleet of Ohio-class ballistic missile submarines (SSBNs) as they begin to retire in 2027. The Navy began technology development in January 2011 in order to avoid a gap in sea-based nuclear deterrence between the Ohio class's retirement and the production of a replacement. The Navy is working with the United Kingdom to develop a common missile compartment for use on OR and the United Kingdom's replacement SSBN. OR will initially carry the Trident II D5LE missile.

Current Status of OR

The OR program is pursuing design for affordability initiatives and investigating various contracting and acquisition scenarios to reduce average follow-on ship procurement costs from an estimated $5.6 billion to $4.9 billion (in fiscal year 2010 dollars). The program intends to maximize economic order of quantity benefits by leveraging the *Virginia*-class program and the elements common with the United Kingdom's SSBN. Program officials stated the OR and *Virginia*-class programs are aligned to support various contracting scenarios, including procuring the lead OR concurrently in a multiyear procurement with the Block V *Virginia*-class contract. The Navy will develop a legislative proposal in 2017 to support this approach.

The Navy has set initial configurations for areas including the torpedo room, bow, and stern. In 2014, the program completed ship specifications, set ship length—a major milestone—and started detailed system descriptions and arrangements. The contractor plans to build a notional submarine section to validate the fidelity of its new design tool. The program plans to have 83 percent of design disclosures (including drawings, and material procurement and construction planning information) and 100 percent of arrangements and the 3D product model completed prior to the start of lead ship construction—scheduled for fiscal year 2021. The Navy Capability Definition Document (CDD) was approved August 2012. The program is aligned with the JCID manual to submit to the Joint Requirements Oversight Council (JROC) for approval to support Milestone B in August 2016.

the program office, it requires support from financial management, contracting, personnel, and potentially other communities requiring leadership at the Chief of Staff and Secretary of the Air Force levels to make this a priority on critical acquisition programs. Several participants noted that many presentations cited the need for barriers to be lifted in the financial management community (using 3600 funding to resource an increase in the number and improve the technical capabilities of the acquisition workforce), contracting community (providing access to higher technical talent subject-matter experts for high-quality engineering services support, for example), and personnel community (allowing movement of high-quality organic personnel across programs within a PEO's UMD for example). It is clear to some that these barriers and limitations need to be identified and actions taken in multiple functional communities to enable programs to meet any objectives set for owning the technical baseline. However, other participants believe ownership of the technical baseline should be deliberate on the programs that require it, but not on those that do not. They added that a future or follow-on study involving members of this functional community to jointly look for barriers and solutions to lift them would help to ensure a unified and cohesive Air Force effort.

Appendixes

A

Biographical Sketches of Committee Members

HENRY A. OBERING III, *Chair*, is a Booz Allen Hamilton executive vice president. An expert in acquisition and program management, he works with aerospace clients in the Air Force and National Aeronautics and Space Administration (NASA). Prior to joining Booz Allen Hamilton, he led a comprehensive review of the National Reconnaissance Office (NRO) for the Director of National Intelligence that provided a new charter for that organization. Mr. Obering retired from the U.S. Air Force as a lieutenant general with more than 35 years of experience in space and defense systems development, integration, and operations. He served as director of the 8,500-person Missile Defense Agency (MDA). He was the Department of Defense (DoD) acquisition executive for the nation's $10 billion per year missile defense portfolio. In addition, he served as the program manager for the Ballistic Missile Defense System. Prior to his assignment at MDA, he planned and programmed 68 joint Air Force and international programs with a $28 billion budget as Mission Area Director for Information Dominance on the Air Staff. Mr. Obering entered the Air Force in 1973 after completing the University of Notre Dame's ROTC program as a distinguished graduate. He received his pilot wings in 1975 and flew operational assignments in the F-4E. Later, he was assigned to the Space Shuttle Program and participated in 15 space shuttle launches as a NASA orbiter project engineer and was responsible for integrating firing room launch operations. Other assignments include tours with the Air Force Inspector General, the Defense Mapping Agency, and Electronic Systems Center. Mr. Obering has twice earned the DoD's highest non-combat award, the Defense Distinguished Service Medal for leadership. He was honored by the National Defense Industrial Association's Missile Defense Division with the 2011 Kadish Award for Acquisition Excellence. He received an M.S. in astronautical engineering from Stanford University.

CLAUDE M. BOLTON, JR., became the executive-in-residence for the Defense Acquisition University (DAU) in January 2008. His primary focus is assisting the DAU president achieve the congressional direction to recruit, retain, train, and educate the DoD acquisition workforce. Mr. Bolton is also a management consultant to defense and commercial companies and is a board member for several companies. Prior to becoming the DAU executive-in-residence, He served as the Assistant Secretary of the Army for Acquisition, Logistics, and Technology where he served as the Army acquisition executive, the senior procurement executive, and the science adviser to the Secretary. Mr. Bolton oversaw the Elimination of Chemical Weapons Program, and had oversight and executive authority over the Project and Contracting Office charged with Iraq reconstruction. He was responsible for appointing, managing, and evaluating program executive officers as well as managing the Army Acquisition Corps and Army Acquisition Workforce. Mr. Bolton retired as a major general in the U.S. Air Force following a highly decorated career. Some highlights of Mr. Bolton's Air Force service include serving as the Commander, Air Force Security Assistance Center, where he managed foreign military sales programs with totals exceeding $90 billion that supported more than 80 foreign countries; serving as a test pilot for the F-4, F-111, and F-16; program executive officer for the Air Force Fighter and Bomber programs; and the first program manager for the Advance Tactical Fighter Technologies program, which evolved into the F-22 System Program Office. An experienced command pilot flying more than 40 different aircraft including Army helicopters; during the Vietnam War he flew 232 combat missions, 40 over North Vietnam. Mr. Bolton served as commandant of the Defense Systems Management College, and as inspector general and director of requirements at Air Force Materiel Command (AFMC) headquarters. He holds an M.S. in management from Troy State University and an M.A. in national security and strategic studies from the Naval War College. In 2006, he was awarded a D.Sc. (honoris causa) from Cranfield University. In 2007, he was awarded an honorary doctor of science degree from the University of Nebraska, Lincoln, his alma mater. Mr. Bolton is a member of the National Research Council's (NRC's) Air Force Studies Board and is a past member of the NRC Committee on Evaluation of U.S. Air Force Preacquisition Technology Development and the Committee on Optimizing U.S. Air Force and Department of Defense Review of Air Force Acquisition Programs.

MICHAEL D. GRIFFIN is chairman and CEO of Schafer Corporation, a leading provider of scientific, engineering, and technical services and products in the national security sector. He was previously King-McDonald Eminent Scholar and professor of mechanical and aerospace engineering at the University of Alabama, Huntsville. He was the Administrator of NASA from 2005-2009 and prior to that was the Space Department Head at the Johns Hopkins University Applied Physics Laboratory (JHU/APL). He has also held numerous executive positions with industry, including president and chief operating officer of In-Q-Tel, CEO of Magellan Systems, general manager of Orbital Science Corporation's Space Systems Group, and executive vice president and chief technical officer at Orbital. Dr. Griffin's earlier career includes service as both chief engineer and associate administrator for exploration at NASA and as the deputy for technology at the Strategic Defense Initiative Organization (SDIO). Prior to joining SDIO in an executive capacity, he played a key role in conceiving and directing several "first of a kind" space tests in support of strategic defense research, development, and flight testing. These included the first space-to-space intercept of a ballistic missile in powered flight, the first broad-spectrum spaceborne reconnaissance of targets and decoys in midcourse flight, and the first space-to-ground reconnaissance of ballistic missiles during the boost phase. He also played a leading role in other space missions in earlier work at JHU/APL, NASA's Jet Propulsion Laboratory, and the Computer Science Corporation. Dr. Griffin was an adjunct professor for 13 years at the University of Maryland, JHU, and George Washington University, offering courses in spacecraft design, applied mathematics, guidance and navigation, compressible flow, computational fluid dynamics, spacecraft attitude control, astrodynamics, and introductory aerospace engineering. He is a registered professional engineer in Maryland and California and is the lead author of over two dozen technical papers and the textbook *Space Vehicle Design*. Dr. Griffin is a member of the National Academy of Engineering and the International Academy of Astronautics, an honorary fellow and the current president of AIAA, a fellow of the American Astronautical Society, and a senior member of the Institute of Electrical and Electronic Engineers. He is the recipient of numerous honors and awards, including the NASA Exceptional Achievement Medal, the AIAA Space Systems Medal and Goddard Astronautics Award, the National Space Club's Goddard Trophy, the Rotary National Award for Space Achievement, the Missile Defense Agency's Ronald Reagan Award, and the DoD Distinguished Public Service Medal, the highest award that can be conferred on a non-government employee. He received his Ph.D. in aerospace engineering from the University of Maryland and has been recognized with honorary doctoral degrees from Florida Southern College and the University of Notre Dame.

TERRY J. JAGGERS is a former chief engineer of the U.S. Air Force and former deputy chief engineer of DoD. He graduated from the Industrial College of the Armed Forces with a master's degree in national security, the Florida Institute of Technology with a master's degree in business administration, the University of Illinois with a bachelor's degree in aerospace engineering, and Western Illinois University with a bachelor's degree in mathematics. He holds appointments as senior research fellow at the University of Texas, Austin, and associate fellowship at AIAA. He serves on numerous professional societies and advisory boards including the University of Illinois

Aerospace Engineering Department, the University of Maryland System Research Institute, and the Science and Technology Policy Institute, where he provides science, technology, engineering, and STEM workforce advice to the White House Office of Science and Technology Policy and various agency S&T executives. Mr. Jaggers is currently chief scientist at Decisive Analytics Corporation, where he provides strategic advice to the CEO on the future of systems engineering, big data analytics, intelligence, surveillance and reconnaissance fusion for missile defense and space, and cyber protection. Prior to this, he was the NRC board director for both U.S. Air Force and Office of Director of National Intelligence programs, overseeing $45 million in studies and recommendations to senior leaders spanning science and technology, development planning system acquisition, space protection, strategic deterrence, sustainment, and other critical national security issues. He is a former Principal Deputy Assistant Secretary of Defense for Systems Engineering and former Deputy Assistant Secretary of the Air Force for Science, Technology, and Engineering where he provided program engineering and technical management policy for over $60 billion in major acquisition category I and II programs, functional development for more than 40,000 scientist and engineer professionals, and oversight for the $2 billion Air Force S&T program as S&T executive. He was directly responsible for instituting the early systems engineering initiative leading to the enactment of the Weapon Systems Acquisition Reform Act and subsequent revitalization of development planning, launched the first-ever Air Force STEM Strategic Plan leading to the Air Force STEM Advisory Council, and is a member of the national security space workforce cadre serving in launch and range operations, development planning, and joint space program office assignments.

GARY A. KYLE is president and CEO of Persistent Agility, Inc., a veteran-owned small business providing unique, innovative, and proven business and mission solutions. He leads and participates on senior-level teams that independently advise C-level executives and U.S. government senior leaders on complex multibillion dollar acquisition, contracting, and strategy matters. Mr. Kyle served as a principal at Booz Allen Hamilton, where he led a team of 52 business analytics consultants providing acquisition, contracting, cost, and financial expertise for Air Force clients. He was hand selected by the Air Force Space Command and the NRO as the business lead for the Evolved Expendable Launch Vehicle Tiger Team that defined the program's future acquisition tenets. As the Verizon senior vice president for global military sales, Mr. Kyle drove the organization's record revenue growth. He also led the staff responsible for financial management and customer service for the $900+ million defense and international services organization. Mr. Kyle served as a senior Air Force officer and held high-visibility assignments both in the Pentagon and supporting multibillion dollar major defense acquisition programs. Mr. Kyle received numerous team and personal awards, including the DoD David Packard Excellence in Acquisition Award; DoD Value Engineering Award; DoD Defense Standardization Program National Honorary Award; Air Force John J. Welch, Jr. Award for Excellence in Acquisition Management (twice); Secretary of the Air Force Contracting Award for Outstanding Strategic Acquisition Reform; and Los Angeles Chapter of the Air Force Association Officer of the Year Award. Mr. Kyle holds an M.A. in telecommunications from the George Washington University, an M.S. in national resource strategy from National Defense University, and an M.S. in procurement and acquisition management from Northrop University. He is a graduate of the Defense Systems Management College Program Management Course and holds the highest DoD acquisition certifications in both program management and contracting.

JON S. OGG is CEO and principal partner at AeroVise LLC. Mr. Ogg retired in 2008 after 33 years with the Air Force serving in multiple assignments of increasing responsibility spanning defense systems research, development, and sustainment. His clients include the Air Force, Defense Advanced Research Projects Agency, Defense Threat Reduction Agency, DoD, and various defense companies. Mr. Ogg is an experienced systems engineer who spent his first 15 years tackling engineering challenges spanning virtually all Air Force fighter, bomber, airlift, and trainer propulsion systems in the field today. In 1991, he expanded his breadth of experience and responsibility to a critical leadership role serving as chief engineer for the F-22 Advanced Tactical Fighter (ATF) for nearly a decade, bringing it from source selection to production. Mr. Ogg served as chief/lead engineer for multiple propulsion programs, including F100, F110, TF-34, F101, YF120, YF119, and Cruise Missile systems. He led the development and initial production of F-22 ATF while acting as a primary interface to senior Air Force and Office of the Secretary of Defense personnel. He took the lead in revitalizing systems engineering across AFMC and stood up a classified development planning organization responsible for application of advanced technologies in birthing new weapon systems. As the director of engineering for the research, acquisition, and sustainment arm of the Air Force, Mr. Ogg was responsible for a technical workforce of more than 12,000 and the processes and policies governing their operations. He led numerous high-level teams spanning acquisition strategies, affordability assessments for Global Hawk, T-6 Trainer, and F-22, request for proposal preparations, risk assessments, flight certifications, counter-IED solutions, and generic production cost-cutting initiatives. Mr. Ogg was the chief information officer for AFMC. In this capacity, he was the command's primary catalyst for e-business transformation, cutting more than $200 million (20 percent) out of operations during his 2.5-year tenure. Mr. Ogg established direct links between 6.2/6.3 research and rapid fielding of counter-IED, advanced turbine engine, and hydrocarbon-based hypersonic propulsion. He institutionalized modeling and simulation and systems-of-systems focus as key enablers for reducing development cycle time while providing integrated solutions to warfighters' needs. Mr. Ogg holds an M.S. in aeromechanical engineering from Ohio State University and an MBA from MIT.

SUE C. PAYTON is president of SCI Aerospace, Inc. She is a former Assistant Secretary of the Air Force for Acquisition and former Deputy Under Secretary of Defense. Ms. Payton served as a presidential appointee in two presidential administrations, with responsibility for acquisition, procurement, and technology transition. In the Air Force, she directed an annual budget in excess of $30 billion that included major weapon systems such as unmanned aircraft, fighter aircraft and munitions, information technology, alternative energy, advanced manufacturing, and intelligence programs. Prior to her government service in 2001, she worked in senior leadership capacities at Lockheed Martin/Martin Marietta with responsibility for leading DoD initiatives in areas such as information technology, complex space systems, intelligence, and operations. Her program achievements have been noted on numerous television shows and dozens of media publications, and she is the inaugural recipient of the DoD 2011 Manufacturing Technology Champion Award. In her role as the president of SCI Aerospace, she serves on industry and research laboratory advisory boards. A graduate of Eastern Illinois University (EIU), she earned master's degrees in systems acquisition management and computer information systems from the University of Southern California and Nova Southeastern University, receiving an honorary Ph.D. in public service from EIU. She is a member of many government-industry/educational/nonprofit organizations, including the EOD Warrior Foundation, U.S. Geospatial Intelligence Foundation, the EIU Foundation board of directors, the Air Force Academy Systems Engineering Program Advisory Council, Purdue University president's council, SMU's Caruth Institute, and the Doolittle Institute.

RICHARD T. ROCA is director emeritus at JHU/APL. He became director at JHU/APL in January 2000 and stepped down from that position in June 2010. While director, he led a not-for-profit University Affiliated Research Center that performs research and development work on behalf of DoD, primarily the U.S. Navy, and for NASA and other federal agencies. The organization has on the order of 5,000 employees, over two-thirds of whom are engineers and scientists. Currently, Dr. Roca is a member of the board of directors of the Charles Stark Draper Laboratory, Inc. He spent the first years of his professional career with the AT&T Corporation. He joined Bell Laboratories to design communications equipment and networks, and rose over three decades to major corporate leadership roles there. Immediately before his move to JHU/APL, Dr. Roca was vice president for IP services planning and development at AT&T Laboratories (the AT&T successor to Bell Labs). Other senior roles included chief technical officer of AT&T Solutions, the professional services business unit of AT&T; Civilian Markets Vice President in AT&T's federal systems business unit; and executive director at AT&T Bell Laboratories of the Special Services Division and the Government Information Systems Division. He is a fellow of the American Society of Mechanical Engineers and a past vice president of its Board on Engineering Education. He is a visitor for the Accreditation Board for Engineering and Technology, which accredits U.S. engineering schools. Dr. Roca received an Sc.D. from MIT in mechanical engineering. In 1977, he was awarded a congressional fellowship and worked on the National Energy Act within the House Commerce Committee's Subcommittee on Energy and Power. In 2011, he received the Secretary of Defense Award for Outstanding Public Service.

B

Workshop Presentations and Participants

SESSION 1
NOVEMBER 4-5, 2014
WASHINGTON, D.C.

Air Force Sponsor Perspectives
David Walker (SES), Deputy Assistant Secretary of the Air Force (Science, Technology, and Engineering)

Expeditionary Combat Support System
Jeff Stanley (SES), Deputy Director, Intelligence, Surveillance, Reconnaissance, and Requirements Directorate, Headquarters Air Force Materiel Command

Air Force Initiatives
Bud Boulter, Office of the Deputy Assistant Secretary of the Air Force (Science, Technology, and Engineering)
David Crawford, Principal Engineer, MITRE

T-X
Kevin Buckley (SES), Program Executive Officer
Col Peter Eide, Program Manager

Air Force Life Cycle Management Center Initiatives
Col Michael Meyer, Deputy Director for Engineering

Long-range Standoff Replacement
Ken Lockwood, Director of Engineering, Strategic Systems Division (AFLCMC/EBB)

Family of Advanced Beyond-Line-of-Sight Terminals
Thomas Kobylarz, Chief Engineer

NOTE: The committee members (Appendix A) participated in all sessions of the workshop.

JSTARS Recapitalization
David Genovese, Director of Engineering, JSTARS Recap Program, Airworthiness Designated Technical Authority (AFLCMC/HBR)

Air Force Initiatives and Perspectives
HON William LaPlante, Assistant Secretary of the Air Force (Acquisition)

OSD Initiatives and Perspectives
HON Katrina McFarland, Assistant Secretary of Defense (Acquisition)

Mr. Naim Awwad, Space and Missile Systems Center, Directorate of Engineering

Ms. Cynthia L. Schurr, SAF/AQRE

SESSION 2
DECEMBER 17-18, 2014
DAYTON, OHIO

AFMC and AFLCMC Initiatives and Perspectives
Susan Thornton (SES), Director of Engineering and Technical Management, Air Force Materiel Command
Jorge Gonzales, (SES), Engineering and Technical Management/Services Directorate, Air Force Life Cycle Management Center

AFMC Initiatives and Perspectives
Steven Zamparelli (SES) (AFMC/PK), Director of Contracting
Tom Robinson (SES) (AFLCMC/PK)
Tom Doyon (SES), Director of AFMC Law Office

Tanker Directorate (AFLCMC/WK)
Brig Gen Duke Richardson

Fighter-Bomber Directorate (AFLCMC/WW)
Col Robert Strasser, System Program Manager, B-2 Division

ISR and SOF Directorate (AFLCMC/WI)
Doug Atkinson, Chief Engineer, AFLCMC/WI
Dave Schairbaum, System Program Director, Combat Rescue Helicopter

AFLCMC Initiatives and Perspectives
Lt Gen John Thompson, Commander, Air Force Life Cycle Management Center

DCGS (AFLCMC/HB)
Col Raymond Wier, Chief, C2ISR Division

Dr. David E. Walker (SES), Deputy Assistant Secretary of the Air Force for Science, Technology, and Engineering, Office of the Assistant Secretary of the Air Force for Acquisition, Washington, D.C.

Col Chris Coombs, KC-46 System Program Manager, USAF

Col Douglas S. Martin, Reserve Adviser, DAS ST&E, SAF/AQR

Col Mike Meyer, Deputy Director, Air Force Life Cycle Management Center/EN-EZ

Col Chip Mosle, Chief, KC-46 Development and Production, USAF

Maj Brian Scozzaro, Tanker Directorate Executive Officer, USAF

Cap Jennifer Condon-Pracht, Chief, Tanker Program Execution Group, USAF

Mr. Naim Awwad, Systems Engineering, SMC/EN

Mr. Joe Baker, KC-46 Development Chief Engineer, USAF

Dr. Kenneth Barker, Senior Leader. Air Force Life Cycle Management Center/EN-EZ

Mr. Luke Burke, Director of Engineering, USAF

Mr. Chris Fall, National Security and International Affairs, White House OSTP

Ms. Mary Beth Kundert, Chief, Configuration Management, USAF

Mr. Don Thompson, Deputy Air Force Program Executive Officer for Tankers, USAF

Mr. Larry Rogers, KC-46 Chief Engineer, USAF

Mr. Duane L. Sevey, AFLCMC/WIH, Combat Rescue Helicopter

Mr. John Slye, Tanker Directorate Chief Engineering, USAF

Mr. Bert Turner, Deputy, KC-46 System Program Manager, USAF

Philomena M. Zimmerman, Deputy Director, Engineering Tools and Environments, OSD (AT&L)/OASD (R&E)/ODASD (SE)

SESSION 3
JANUARY 20-22, 2015
WASHINGTON, D.C.

AEGIS
RADM Kathleen Paige (USN, Ret.), Former Director, AEGIS Ballistic Missile Defense
Don Mitchell, Member of the Principal Professional Staff, Johns Hopkins University Applied Physics Laboratory

F-35
Lorraine Martin, F-35 Executive Vice President and General Manager, Lockheed Martin

National Reconnaissance Office
Tina Harrington, Director, SIGINT Directorate

Professional Services Council
HON Stan Soloway, President and Chief Executive Officer

Joint Light Tactical Vehicle
COL John Cavedo, Jr., Program Manager, Combat Support and Combat Service Support, U.S. Army

Combat and Mission Support
Robert Clarke, Deputy Program Executive Officer, Office of the Assistant Secretary of the Air Force (Acquisition)

Launch Test Range System
Col Janet Grondin, Chief, Range and Network Division (SMC/RN), Air Force Space and Missile Systems Center

Command and Control System-Consolidated (CCS-C) Advanced Extremely-High Frequency
John Morris, Chief Engineer, MILSTATCOM Systems Directorate, Air Force Space and Missile Systems Center

F-35
Lt Gen Christopher Bogdan, U.S. Air Force, Program Executive Officer

Ohio Replacement
RADM Dave Johnson, PEO Submarines

Strategic Systems Programs
VADM Terry Benedict, Director

Air Force Life Cycle Management Center
Lt Gen John Thompson, Commander

Dr. David E. Walker (SES), Deputy Assistant Secretary of the Air Force for Science, Technology, and Engineering, Office of the Assistant Secretary of the Air Force for Acquisition, Washington, D.C.

Lt Col David Buchanan, Executive Officer, F-35 Joint Program Office

Capt Adam Brown, Executive Officer to Director, SIGINT Systems Acquisition, National Reconnaissance Office

Mr. Naim Awwad, Systems Engineering, SMC/EN

Mr. Kevin Keck, AZS Division Chief, USAF

Ms. Lynn Eviston, Director, Plans and Programs Directorate, USAF

Mr. Brian Laird, Chief Engineer, Launch Test Range System (SMC/RNSE), USAF

Mr. Peter Manternach, Deputy Chief Engineer, Joint Program Office, Joint Light Tactical Vehicles

1st Lt Rachel Milliron, Aide-de-Camp to the AFLCMC Commander, USAF

Mr. Rocky Reiners, SAF/AQR

Ms. Cindy Schurr, SAF/AQR

Lt Col Mike Taraborelli, Chief, Range Modernization Branch (SMC/RNG), USAF

C

Suggested Terms of Reference for a Follow-on Study

This appendix addresses Task 3 of the committee's terms of reference.

The National Research Council (NRC) will appoint a study committee to conduct a consensus study in accordance with NRC procedures. The NRC will then:

1. Identify the strategic value to the Air Force in properly controlling the technical baselines of its programs.
2. Investigate how others (e.g., Services, government agencies, and commercial industry) control technical baselines and measure success.
3. Recommend ways to remove or remediate the barriers across the Air Force, such as those identified in the workshop, to properly control technical baselines.
4. Recommend ways to resource and institutionalize proper control of the technical baselines, to include addressing cultural factors.

D

Organization of Air Force Acquisition Centers

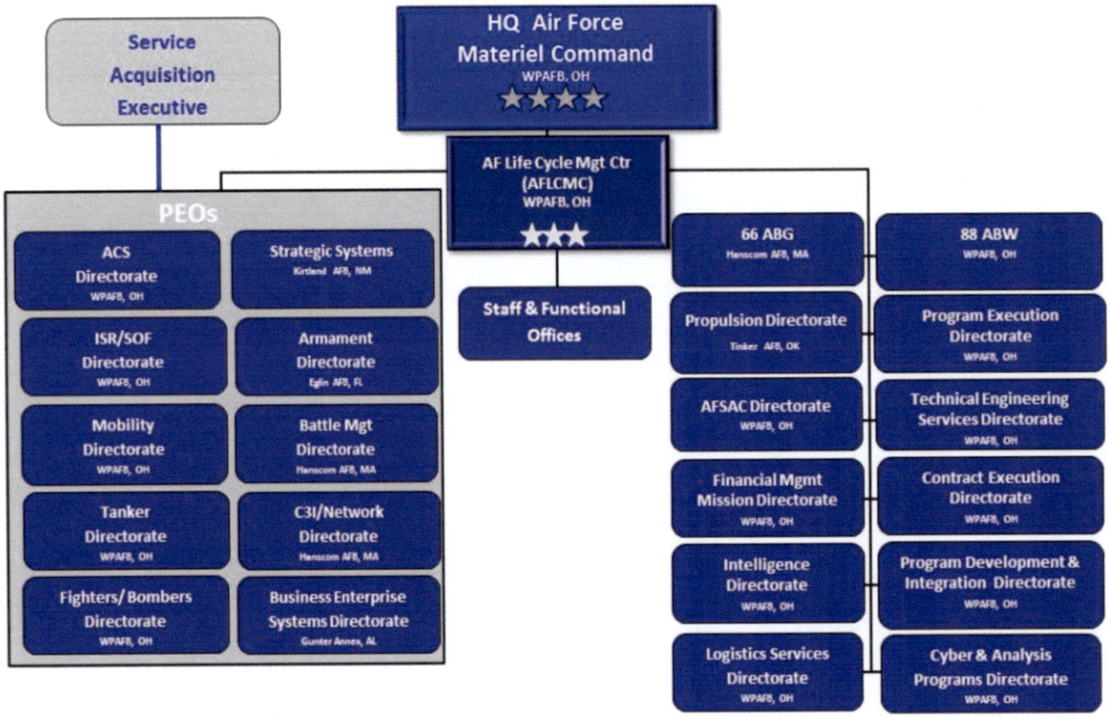

FIGURE D-1 Organization of the Air Force Life Cycle Management Center. SOURCE: U.S. Air Force.

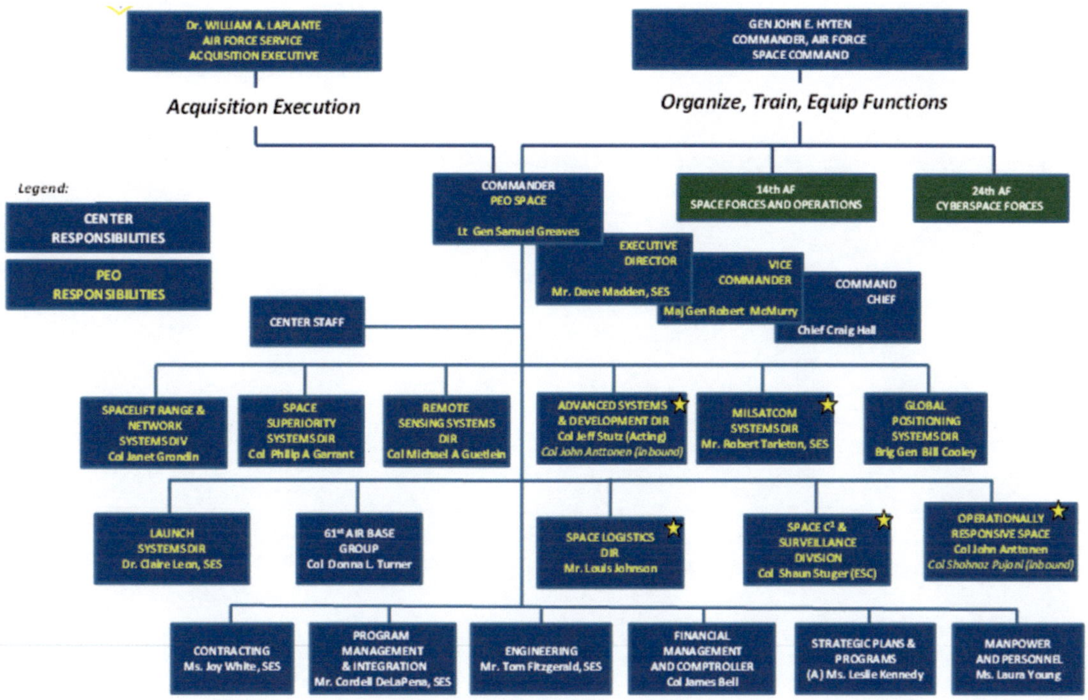

FIGURE D-2 Organization of the Air Force Space and Missile Systems Center. SOURCE: Col Jeff Dickson, Deputy Chief of Engineering, SMC/EN.